Structural Chemistry

STRUCTURAL CHEMISTRY

Dr. Rose Marie O. Mendoza

ARCLER PRESS

www.arclerpress.com

Structural Chemistry

Dr. Rose Marie O. Mendoza

Arcler Press
224 Shoreacres Road
Burlington, ON L7L 2H2
Canada
www.arclerpress.com
Email: orders@arclereducation.com

ISBN: 978-1-77469-827-3 (Hardcover)

Arcler Press publishes wide variety of books and eBooks. For more information about Arcler Press and its products, visit our website at www.arclerpress.com

ABOUT THE AUTHOR

Dr. Rose obtained her PhD in Chemical Engineering from the University of the Philippines-Diliman in 2013, while her Masters in Chemical Engineering and BS Chemical Engineering degree was obtained from Adamson University. She is also a Professor in the Graduate School Department under the Master of Engineering Program at Adamson University since 2006. In 2017, she obtained her Post Doctorate Degree in Green Power: Hydrogen Generation and Fuel Cell Development from the University of California Merced, USA (in collaboration with the Energy Storage and Conversion Materials Laboratory of the University of the Philippines, Diliman). She is a visiting scientist to several universities abroad such as the University of California Merced in USA, Kindai University in Japan, National Taiwan University and Chia Nan University of Pharmacy and Science in Taiwan. She has published more than 20 research papers in water treatment and remediation, green technology, fuel cells, electrodialysis, nanomaterials, high voltage spatial electricity, batteries in peer-reviewed journals, and have more than 15 textbooks published in chemical engineering and processes. As an educator, she is very active in publishing her research to local peer reviewed journals. An advocate of Climate Action, a Climate Reality Leader Corp Member and a budding Sustainability and Circularity talent, Dr. Mendoza is also presently actively engaged in several local and international Green Federation as Board Member and Officer such as the Green Party of the Philippines, and Treasurer of the Asian Pacific Federation of Green Women's Network. She is also one of the Board of Directors of the Philippine Society of Engineering Educators, Mark Energy Revolution Corporation and one of the pioneering members of the Philippine Institute of Chemical Engineers, CaMaNaVa Chapter, and an Associate Member of the National Research Council of the Philippines (NRCP).

TABLE OF CONTENTS

List of Figures .... *xi*

List of Tables .... *xv*

List of Abbreviations .... *xvii*

Preface .... *xix*

Chapter 1 Introduction to Structural Chemistry 1

Unit Introduction 2

1.1. Chemical Structures 3

1.2. Chemical Bonds 10

1.3. Types of Bonds: Covalent, Ionic, Metallic 11

1.4. Ionic Bond 12

1.5. Covalent Bond 13

1.6. Metallic Bond 15

1.7. Summary 18

Review Questions: 18

References 19

Chapter 2 Intermolecular Forces 27

Unit Introduction 28

2.1. Van Der Waals Interaction 28

2.2. Interdependence of the Lengths of Van Der Waals and Covalent Bonds 35

2.3. Van Der Waals Radii 37

2.4. Interactions of Donor–Acceptor 45

2.5. Hydrogen Bond 46

2.6. Summary 50

References 51

Chapter 3 Polymorphism and Phase Transition **59**

3.1. Thermodynamic Stability 60

3.2. Kinetic Stability 60

3.3. Polymorphism 61

3.4. Energies of Phase Transitions 68

3.5. Melting Heats of Compounds 69

3.6. Sublimation Heats of Elements and Compounds 70

3.7. Evaporation Heats of Compounds 73

3.8. Enthalpies of Phase Transformations 75

3.9. Summary 76

Review Questions: 77

References 78

Chapter 4 Properties of Organic Molecules **85**

Unit Introduction 86

4.1. Organic Molecules 86

4.2. Functional Groups 89

4.3. Lewis Structure of Organic Molecules 95

4.4. Hybridization 96

4.5. Acid Base Reactions and K_a/Pk_a Values 102

4.6. Summary 106

Review Questions: 107

References 108

Chapter 5 Structure and Properties of Inorganic Chemistry **115**

Unit Introduction 116

5.1. Introduction 116

5.2. Crystal Structures of Inorganic Compounds 119

5.3. Chemical Bonding 123

5.4. Ionic Bond 124

5.5. Covalent Bond 128

5.6. Physical Properties of Inorganic Compounds 133

5.7. Application of Inorganic Compounds 135

5.8. Summary 136

Review Questions: 137
References 138

Chapter 6 Structure and Optical Properties 145
Unit Introduction 146
6.1. Refractive Index (RI) 147
6.2. Refractive Indices: Effect of Composition, Structures, and Thermodynamic Conditions 152
6.3. Dipole Moments and Polarization 155
6.4. Szigeti Equation 157
6.5. Structural Application of Refractometry 160
6.6. Summary 163
Review Questions: 164
References 165

Chapter 7 The Element Structure of Non-Metals 171
Unit Introduction 172
7.1. Hydrogen and the Halogens 172
7.2. Chalcogens 173
7.3. Elements of the Fifth Main Group 176
7.4. Elements of the Fifth and Sixth Main Groups Under Pressure 179
7.5. Carbon 181
7.6. Boron 185
7.7. Summary 187
Review Questions: 187
Multiple Choice Questions: 188
References 189

Chapter 8 Symmetry as the Organizing Principle for Crystal Structures 193
Unit Introduction 194
8.1. Crystallographic Group–Subgroup Relations 196
8.2. The Symmetry Principle in Crystal Chemistry 199
8.3. Structural Relationships By Group–Subgroup Relations 200
8.4. Diamond–Zinc Blend 202
8.5. The Relation Between Nias and MNP 203
8.6. Symmetry Relations at Phase Transitions: Twinned Crystals 207

8.7. Summary 211
Review Questions 212
References 213

Index 219

LIST OF FIGURES

Figure 1.1. A view of ellipsoids of thermal motion

Figure 1.2. Coordination polyhedral

Figure 1.3. Primitive cubic crystal lattice

Figure 1.4. Tetrahedron of chemical bond types

Figure 1.5. Potential energy (U) of the bonding and antibonding orbitals of a diatomic molecule as functions of the interatomic distance R

Figure 2.1. Determination of distances of Van der Waals using the Morse potential curve

Figure 2.2. Interactions between molecules according to Allinger

Figure 2.3. Interdependence of the lengths of Van der Waals as well as covalent bonds

Figure 2.4. Distribution of intermolecular distances

Figure 2.5. An A2 molecule shown as overlapping spheres

Figure 2.6. Calculation of the vdW radius of M in molecules MX

Figure 2.7. Anisotropic vdW radii

Figure 2.8. Relations of the H-atom with atoms of other elements

Figure 3.1. An example of reconstructive phase transition

Figure 3.2. An example of displacive transition

Figure 3.3. Sublimation heats of elementary solids, periods 1–5

Figure 4.1. The structure of the alkanes is built on a carbon chain, and each carbon atom makes four covalent bonds

Figure 4.2. Strong covalent bonds and weaker intermolecular interactions holding together the atoms in molecules

Figure 4.3. A model of a short segment of the basic polymer poly(ethene) molecule. Numerous chains made up of thousands of carbon atoms make up polyethene molecules

Figure 4.4. Alkane, alkene, and alkyne examples with emphasis on their functional groups. They are all hydrocarbon compounds

Figure 4.5. The hydroxyl group and a halogen atom are examples of functional groups that can be found in alcohols and halogenoalkanes, respectively

Figure 4.6. The carbonyl group is an illustration of a functional group present in aldehydes and ketones

Figure 4.7. The carboxyl group, acyl chloride group, and ester link are examples of functional groups that can be found in carboxylic acids, acyl chlorides, and esters, respectively

Figure 4.8. The amine group and the amide group are examples of functional groups in amines and amides

Figure 4.9. Arenes, haloarenes, and phenols are a few examples of functional groups. An aromatic arene ring can be found in all three families. In contrast to phenols, which have a phenyl group attached to the hydroxyl group, haloarenes are composed of a phenol group connected to a halogen atom

Figure 4.10. The remaining p orbital is used for the π bond

Figure 4.11. The remaining two p orbitals, form 2 perpendicular bonds

Figure 4.12. Tetrahedral and sp^3 hybridized carbons are both present in ethane. Ethane can exist in a variety of orientations by rotating one of the end methyl groups

Figure 4.13. Sigma and Pi bond in ethylene

Figure 4.14. Pi bond break upon rotation of CH_2

Figure 4.15. -cis-but-2-ene and trans-but-2-ene

Figure 4.16. Structural isomers of n-butane and isobutane

Figure 4.17. Structural isomers of C_5H_{10}

Figure 4.18. Stereoisomers and structural isomers

Figure 5.1. (a) A (UC15-2) molecule is depicted graphically in identical scales. Valence-bond formula, in. (b) An elliptical perspective picture of thermal motion. (c) Polyhedra for coordination. And (d) emphasis on chlorine atom's need for space

Figure 5.2. Primitive lattice of cubic crystals. A unit cell

Figure 5.3. Atoms A and X are positioned in a regular, two-dimensional pattern. Any one of the depicted unit cells can be repeated to create the entire pattern

Figure 5.4. Primitive and centered cell

Figure 5.5. Unit cells that are centered and their symbols. Numbers indicate how numerously primitive each is

Figure 5.6. For face-centered unit cell of NaCl, here is how to count the contents: 12 Na ions in the centers of 12 edges, each of which belongs to 4 cells, makes 124 3; 1 Na ion in the center of the cube, which belongs only to this cell, makes 88 1. There are also 6 Cl ions in the centers of 6 faces, each belonging to two adjacent cells. Total: 4 units of the formula NaCl (Z = 4) or 4 Na and 4 Cl ions

Figure 5.7. (a) To prevent interaction with air or water, sodium, a soft metal, must be kept in mineral oil while not in use. (b) Chlorine gas has a vivid yellow-green color. (c) They combine to create white crystals of sodium chloride or table salt

Figure 5.8. A sodium chloride atom in table salt are positioned to (a) increase the interaction of opposing charges. While the larger spheres stand in for chloride ions, smaller spheres represent sodium ions. The geometry is easier to see in the larger view.

(b) Keep in mind that each surrounding ion—in this case, six—is "bonded" to every other ion

Figure 5.9. A potential energy diagram showing the interaction of two H atoms. The internuclear bond distance determines the lowest potential energy

Figure 5.10. Lewis acid-base complex formation

Figure 5.11. Transition metal coordination complex representations

Figure 5.12. Tetracarbonyl nickel

Figure 5.13. The metallic bonding in sodium

Figure 6.1. Optical surfaces of both the ordinary (o) and extraordinary (e) beams in mono-axial crystal and also the optically positive (left, $v_o>v_e$) and if the optically negative (right, $v_e>v_o$). AA is the optical axis

Figure 6.2. Illustration of optical indicatrices of mono-axial crystals: + for $n_e> n_o$, – for $n_o> n_e$

Figure 6.3. Refractive indices (n∞) of crystalline alkali halides depends on the melting temperature

Figure 6.4. IR absorption of a single crystal shown in (solid line) and powder shown in (dotted line)

Figure 7.1. Section of the structure of **α-sulfur**

Figure 7.2. Structure of **α-selenium**

Figure 7.3. The structure of black phosphorus

Figure 7.4. Section of a layer in grey arsenic

Figure 7.5. Four body-centered unit cells of the incommensurately modulated structure of tellurium-III

Figure 7.6. The incommensurate structure of Bismuth-III

Figure 7.7. The structure of graphite (stereo-image)

Figure 7.8. Top: Molecular structure of C_{60} molecule (Stereo view). Bottom: Packing of C_{60} molecules and K ions in K_3C_{60}

Figure 7.9. Structures of two kinds of mono-layered carbon nanotubes. The left one shows metallic conduction, the right one is semi-conducting

Figure 7.10. Structures of rhombohedral **α**-B_{12}

Figure 8.1. Examples for translationengleiche group–subgroup relations. Left: loss of reflection planes; right: reduction of the multiplicity of a rotation axis from 4 to 2. The circles of the same type, Æ and, designate symmetry-equivalent positions

Figure 8.2. Examples of group-subgroup relationships that are klassengleiche. Right: Enlargement of the unit cell with symmetrical reduction of reflection planes perpendicular to b to glide planes. Left: Loss of centering, including loss of glide planes and twofold axes. Similar-type circles, such as and, designate symmetry-equivalent places

Figure 8.3. The connection between MnP and NiAs. The numbers represent z coordinates for the photos

Figure 8.4. Sections of the hexagonal closest-packing. Gray area: Unit cell, space group P 63 m 2 m 2 c. Large cell: Base of the triple cell. The z coordinates of the spheres refer to c = c. The dots labeled ①, ②, and ③ mark six octahedral voids at z = 0 and z = 1 (for c = c) and at z = 0 and z = 1 (c = 3c)

Figure 8.5. Group–subgroup relations and the emergence of twins in the phase transition β-quartz α-quartz (Dauphin'e twins)

Figure 8.6. Group–subgroup relations and the emergence of twins in the phase transition β-quartz α-quartz (Dauphin'e twins)

LIST OF TABLES

Table 3.1. Melting heats (KJ/mol) of inorganic compounds

Table 3.2. Sublimation heats of metal oxides (kJ/mol)

Table 3.3. Sublimation (upperlines) and melting heats (lowerlines) of elements (kJ/mol)

Table 3.4. Heats of sublimation of metal chalcogenides (kJ/mol)

Table 3.5. Heats of evaporation of inorganic and organo-metallic compounds (kJ/mol)

Table 3.6. Increasing of the enthalpy of evaporation of organic compounds (kJ/mol)

Table 3.7. Enthalpies of phase transitions of elements (kJ/mol)

Table 4.1. The functional group formula and an example molecule along with the functional group names connected to several organic families

Table 4.2. The pKa table provides details on the breadth and depth of a variety of chemical behavior

Table 7.1. Distances between adjacent atoms and bond angles in structures of the α-As, α-Se, α-Po and β-Po Type

Table 8.1. Crystal data of structures mentioned in Figure 8.4

LIST OF ABBREVIATIONS

AIMs	Atoms in Molecules
IR	Infrared
MC	Mobility Coefficient
ND	Neutron Diffraction
NH_3	Ammonia
NMR	Nuclear Magnetic Resonance
RI	Refractive Index
VB	Valence-Bond
VSEPR	Valence Shell Electron Pair Repulsion Theory
ZPE	Zero-Point Energy

PREFACE

Chemical structures serve as the fundamental building blocks of all matter. They determine substances' properties, reactivity, and behavior, playing a crucial role in fields as diverse as medicine, materials science, environmental science, and biochemistry. By understanding the structure of molecules, we gain insights into their physical and chemical properties, enabling us to design novel materials, develop efficient drugs, and comprehend the intricacies of biochemical processes.

In this book, we explore structural chemistry from its fundamental concepts to its advanced applications. We introduce the basic principles of chemical bonding, molecular geometry, and electronic structure. With a strong emphasis on visual representations, we guide you through the world of Lewis structures, hybridization, VSEPR theory, and molecular orbitals, unraveling the intricate interplay between atoms and electrons.

Moving forward, we delve into various chemical bonding phenomena, including ionic, covalent, and metallic bonds. We uncover the factors that influence bond strength, polarity, and stability and explore the intriguing world of intermolecular forces and their role in determining molecular interactions.

One of the cornerstones of structural chemistry is the study of spectroscopy. Through careful spectra analysis, we can decipher the fingerprints of molecules, unravel their structures, and gain a deeper understanding of their electronic and vibrational properties. In this book, we explore various spectroscopic techniques such as infrared (IR) spectroscopy, UV-Visible spectroscopy, and nuclear magnetic resonance (NMR), equipping you with the tools to interpret and analyze experimental data.

As we progress, we delve into the world of solid-state chemistry, investigating the fascinating structures and properties of crystalline solids. We explore the principles of crystallography, crystal lattice structures, and their impact on the behavior of materials, paving the way for understanding phenomena such as conductivity, magnetism, and phase transitions.

Furthermore, this book sheds light on structural chemistry's pivotal role in elucidating chemical reaction mechanisms. We explore the concepts of reaction kinetics, thermodynamics, and transition state theory, enabling you to unravel the complexities of chemical transformations and their underlying driving forces.

CHAPTER 1

Introduction to Structural Chemistry

Contents

Unit Introduction 2
1.1. Chemical Structures 3
1.2. Chemical Bonds 10
1.3. Types of Bonds: Covalent, Ionic, Metallic 11
1.4. Ionic Bond 12
1.5. Covalent Bond 13
1.6. Metallic Bond 15
1.7. Summary 18
Review Questions: 18
References 19

UNIT INTRODUCTION

The study of chemical structures of compounds, also known as stereochemistry or structural chemistry, is a branch of chemistry that focuses primarily on molecule structures. In structural chemistry, the spatial arrangement of atoms in a compound is clarified and described, along with the causes of this arrangement and which dictates its properties and characteristics. Additionally, it involves the systematic grouping of known structure kinds and the revelation of connections between them (Batsanov & Batsanov, 2012). Structural chemistry is a crucial component of contemporary chemistry in both theory and application. Knowing the molecular makeup of the compounds present is crucial for understanding the processes occurring in a chemical reaction and making it feasible to create studies on new compounds. Only when the structure of a substance is known, its chemical and physical properties can be understood (Lobo, 2003).

When comparing graphite and diamond, it is clear how much a material's structure affects its properties. Despite both being made entirely of carbon, they have very different physical and chemical characteristics (Wells, 2012). The determination of structure is the most significant activity in structural chemistry. It is primarily carried out using X-ray diffraction from single crystals or using diffraction of neutron from single crystals and powders and X-ray diffraction from crystalline powders (Pitzer, 1960). The analytical component of structural chemistry is structure determination, and the typical output is a static model. Experimentally, explaining the spatial rearrangements of atoms that occur during a chemical reaction is much more difficult. In the chemistry of molecules, reaction mechanisms involve this part of structural chemistry (Lockard & Lingrel, 1969).

Remember

Each atom's coordinates (x, y, and z) within the unit cell have values between 0 and 1.

Topotaxy, on the other hand, studies chemical processes in solids where structural relationships are present among the direction of intermediates and products. Real crystals have many flaws, which become more prevalent as the temperature rises. Dislocations and other flaws, such as missing or misplaced atoms, can occur. These flaws have a significant impact on a material's properties (Pyykko, 1988).

1.1. CHEMICAL STRUCTURES

A suitable description of the atoms' spatial distribution is required to determine the molecular makeup of a chemical compound. This is where chemical nomenclature is needed. However, no systematic nomenclature for solid-state structures enables us to describe the details of the structure (Katritzky et al., 2010). For instance, according to the claim that "magnesium fluoride crystallizes in the rutile type," the order of arrangement of Mg and F atoms in MgF_2 corresponds to the distribution of Ti and O atoms in rutile. Each kind of structure is given by an arbitrary representative (Hansch & Fujita, 1964).

The valence-bond (VB) formula makes it possible to sum up the essential structural elements of a molecule. More accurate and illustrative models or true-to-scale figures with the atoms depicted as balls or — if the continuously apparent thermal vibrations are to be said — as ellipsoids (Barger & Dale, 1910). For a clearer view, the balls and ellipsoids are presented at a scale based in their actual atomic sizes. The type of covalent bonds are shown by sticks. It has been found that the size of a thermal ellipsoid corresponds to the chance of discovering the atom, which is typically 50%, averaged over time (Pena & Fierro, 2001) (Figure 1.1).

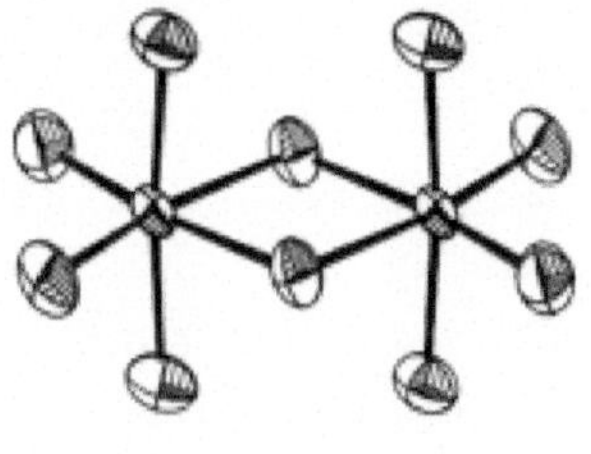

Figure 1.1. A view of ellipsoids of thermal motion.

Source: Wikipedia, Creative Commons License.

Interatomic distances and angles are given numerical values to create quantitative specifications. The distance between two atoms' nuclei in their mean positions is known as the interatomic distance. X-ray diffraction from single crystals is the technique used most frequently to measure interatomic distances experimentally (Liebau, 2012). Other techniques include microwave spectroscopy with gaseous samples, neutron diffraction (ND) from crystals, and electron diffraction for small molecules. The locations of negative charge concentration in the atomic electron shells are that which can be found by X-ray diffraction instead of the positions of the atomic nuclei because atomic electrons diffract X-rays (Duke, 1996). With the exception of hydrogen atoms that are covalently bound, the relative positions of the centers of the negative charge match the atomic nuclei almost exactly. Another factor that makes ND superior to X-ray diffraction in in terms of pinpointing hydrogen atom locations: The location of H atoms in the distribution of heavy atoms can only be found with less reliability due to the large number of heavy atoms' electrons diffracting X-rays to a much greater extent. Neutrons, on the other hand, interact with atomic nuclei. Scattering of neutrons is carried out with deuterated compounds because H atom nuclei inflict incoherent scattering on neutrons to a larger degree as compared to D atom nuclei (Aspinall, 1959).

1.1.1. Coordination Numbers and Coordination Polyhedra

Coordination number and coordination polyhedron is used to describe an atom's immediate environment. The coordination number is the total amount of coordinated atoms (the nearest neighboring atoms). It is easy to state the coordination numbers for every atom in many compounds. However, how far an adjacent atom can be considered as the closest neighbor is not frequently clear (Hoppe, 1970). The number of atoms surrounding each Sb atom in metallic antimony, for instance, is three at 291 pm and an additional three at 336 pm, or only 15% more. In this case, it is useful to express the number used for coordination as 3+3, where the 1st number represents the number of closer-together atoms in the vicinity. In more complex situations, stating an atom's coordination as a single number is not very illuminating (O'Keeffe, 1979). The subsequent specifications can be made, though: an atom has 4 nearby atoms in the white tin at a distance of 302 pm, four at 377 pm and two at 318 pm. Numerous proposals have proposed calculating the mean or "effective" coordination number by combining all the nearby atoms using a weighting scheme, where we count atoms as fractional atoms with a value between 0 and 1, which is closer to 0 as the atom is further away (Ruiz□Martínez & Alvarez, 2009). In the arrangement of the distances between atoms, there is frequently a gap: if the smallest distance between neighboring atoms is 1, then additional atoms are frequently found at distances among 1 and 1.3, after them there is a gap in which no atoms are discovered (Serezhkin et al., 1997). According to a theory proposed by Brunner (1971) and Schwarzenbach (1971), an atom at a distance of

1 ought to have a weight of 1, the first atom outside the gap must possess zero weight, and all of the intermediate atoms must have weights that are calculated via their distances by linear interpolation:

e.c.n. = $\sum i(d_g d_i)/(d_g d_1)$

where; d_1 is the distance to the closest atom; d_g is the distance to the first atom beyond the gap; d_i is the distance to the i-th atom in the area between d1 and dg For instance for antimony: taking 3*d_1 291, 3*d_i 336 and dg 391 pm one finds e.c.n.= 4.65 (Carter, 1978).

When there is no discernible gap, the method is useless. The domain of influence is considered in an original mathematical calculation method. The domain is defined as an array of planes contrary to the lines that connect and pass through the midpoints of all the surrounding atoms, resulting in a convex polyhedron as the domain of influence (Hartshorn, Hey-Hawkins, Kalio, & Leigh, 2007). Every nearby atom can then be given a polyhedron face, with the area of the face acting as a yardstick for the weighting. The largest face is given a value of 1. There have also been other formulas developed. Additionally, the type of bond between adjacent atoms must be considered (Dehnicke & Greiner, 2003). While all atoms 'in contact' are counted, the number of coordinated atoms for a chlorine atom, for instance, in the molecule CCl_4 is 4 (1 C + 3 Cl), as opposed to 1 while just the covalently bonded C atom can be considered. One will typically only count atoms covalently bound together as coordinated atoms in the context of molecules (King, 1996).

In the instance of crystals made up of monoatomic ions, just the cations and anions that are closest to each other are usually taken into consideration, regardless of whether there are connections among anions and anions or among cations (Alvarez et al., 2005). In this way, the coordination number of an I ion in LiI (NaCl) is displayed as 6, but when the 12 other I ions it is in contact with are taken into consideration, the coordination number is 18. When in doubt, it's best to be as specific as possible about what should be a part of the coordination sphere (King, 1970). When the points of simultaneously neighboring coordinated atoms are linked to one another, the coordination polyhedron is formed. There is a typical coordination polyhedron for every coordination number (Marçalo & De Matos, 1989). Even though it may not be apparent at first glance, several coordination polyhedra for an identified coordination number can occasionally be transformed into one another by small atom-level shifts (King, 1969). For instance, by moving four coordinated atoms, a trigonal bipyramid can become a tetragonal pyramid.

Connected polyhedra can be used to describe larger structural elements. A common vertex, face, or edge can be used to join two polyhedra. Bridging atoms are the shared atoms of 2 connected polyhedra. The atoms in the center are nearest to each other in face-sharing polyhedra and farthest apart in vertex-sharing polyhedra (Peresypkina & Blatov, 2000). The notation proposed by Machatschki (1947) (and expanded by a number of additional authors) can be used to represent the coordination conditions in a chemical formula. Next to the symbol for the element, a right superscript contains bracketed information about an atom's coordination number and polyhedron (McGreevy, Baranyai, & Ruff, 1986). The symbols can be expressed in short forms, such as the coordination number by itself and for simple polyhedra, the letter by itself, such as t for the tetrahedron, with the brackets optional in this case (Cotton & Bergman, 1964) (Figure 1.2).

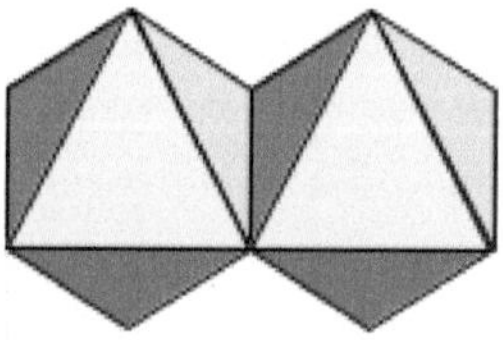

Figure 1.2. Coordination polyhedral.

Source: Wikipedia, Creative Commons License.

1.1.2. Explanation of Crystal Structures

A crystal is a three-dimensional periodic structure made up of a broader network of linked atoms. The positioning of the atoms within space determines the crystal structure. The frequently reiterated atoms of one kind are connected to a three-dimensional form grid across three space directions to form the crystal lattice (Cowan, 1992). Every point in the crystal lattice relates to each other and is enclosed by the same elements, symbolizing a three-dimensional structure of points.

A parallelepiped is a body with six parallel faces, and the crystal lattice can be thought of as being constructed by repeatedly reorganizing a small parallelepiped in each of the three directions with no gaps. The parallelepiped is referred to as the unit cell (Dorrian, Newnham, Smith, & Kay, 1972) (Figure 1.3).

Figure 1.3. Primitive cubic crystal lattice.

Source: Wikipedia, Creative Commons License.

Three basis vectors, indicated through the symbols a, b, and c, together define the unit cell. The set of all linear configurations with t = ua + vb + wc, in which u, v, w is the set of every positive and negative integers, is known as the crystal lattice. The terms "crystal lattice" as well as "crystal structure" should not be used interchangeably as the crystal lattice is a random geometric structure (Glusker & Trueblood, 2010). The lattice variables also referred to as lattice constants (that is, between b and c, etc.), are the dimensions of the basis vectors (a, b, and c), as well as the angles that divide them. The selection of the unit cell over a specific crystal structure does not involve any special techniques (Birle et al., 1968). The selection of the unit cell in crystallography has specific rules that must be followed to achieve uniformity in the creation of crystal structures:

- The basis vectors for the unit cell should be parallel to symmetrical axes or parallel to symmetry planes in order to demonstrate the crystal's symmetry.
- An inversion center is prioritized when choosing a geometrically distinct point to serve as the starting point of the unit cell.
- Use few basis vectors as possible. The cell volume must also be as small as feasible, while the angles among the cells must be as close to 90° as possible.
- The angles among the basis vectors have been selected to be larger or smaller than 90° (which is preferably>90°) if they deviate from that angle (Henrich et al., 2005).

Primitive cells are unit cells with the smallest possible volume. A cell in the center, which is a double, triple, or fourfold primitive, means its volume is bigger by an equivalent factor and is often chosen for symmetry reasons in accordance with rule 1 and contravention of rule 3. To produce standardized explanations of crystal structures, additional rules have been established in addition to those mentioned for cell selection (Lohkamp et al., 2006). They must be attached to each other in order to guarantee methodical and similar records of the data and to make database inclusion easier. Contraventions of the standards, however, are fairly common; they frequently result not only from carelessness or a failure to understand the rules but also from strong justifications, such as when it is necessary to draw attention to the connections between various structures (Stout & Reed, 1954). Just the lattice parameters and the locations of all the atoms of the unit cell are necessary to describe all crucial facets of a crystal structure. Only an integral number of atoms can fit inside of a unit cell (Niebel & Venables, 1974). The quantity of "formula units" per unit cell is provided when stating what is inside the cell; this number is typically abbreviated as Z.

1.1.3. Atomic Coordinates

A collection of atomic coordinates, or the three coordinates x, y, and z, describe an atom's location inside a unit cell. These speak of a coordinate system whose basis vectors are those of the unit cell. The length of each basis vector is equal to the unit length measured alongside the corresponding coordinate axis (McRee, 1999). Thus, every atom's coordinates inside the unit cell have values between 0.0 and <1.0. The system of coordinates is not Cartesian; the coordinate axes may be inclined for each another, and the axes' unit lengths can vary. When an integral number is added to or subtracted from a coordinate value, the coordinates of a corresponding atom in another cell are produced. For instance, the triplet of coordinates $x = 1.27$, $y = 0.52$, and $z = -0.0$ describe the location of an atom within a cell that is next to the origin cell, specifically in the directions +a and –c; this atom is identical to the atom at $x = 0.27$, $y = 0.52$, and $z = 0.90$ in the origin cell. Usually, one asymmetric unit atoms' atomic coordinates are the only ones listed (Svergun, Barberato, & Koch, 1995). The list does not include the atoms that can be "generated" from these by symmetry operations. Pointing out the space group reveals which symmetry procedures are to be used.

All structural information can be inferred from the parameters of lattice, space group, and atomic coordinates (Pauling & Corey, 1951). In particular, it is possible to calculate all distances between atoms and angles.

The corresponding standard deviations, which express the accuracy of their experimental finding, should be provided when defining the coordinates of atoms, distances between atoms, etc. The most common notation, such as "d = 235.1(4) pm," indicates divergence from the mean of 4 units for the final digit, or 0.4 pm in this instance. Statistics has a concept known as standard deviation. The likelihood that the true value will fall within $\pm\sigma$ the bounds of the given value is 68.3% when a standard deviation is associated with some value (Vranka & Amma, 1967). There is a 95.4% chance of being within $\pm 2\sigma$ and a 99.7% chance of being within $\pm 3\sigma$. The standard deviation does not provide accurate information about a value's truthfulness because it only accounts for systematic rather than systematic errors (McLachlan, 1972).

1.1.4. Isotypism

When two compounds have identical atomic distributions and symmetry, their crystal structures are said to be isotypic. When atoms of one element are swapped out for atoms of one more without altering their locations in the crystal structure, one of them can be created from the other. Small changes are allowed for the atomic coordinates, and the lattice dimensions and interatomic distances may have different absolute values (Blatov, 2000). Similar angles are required between the compared lattice dimensions and the crystallographic axes (axes ratios). For all atomic positions, two isotypic structures show a one-to-one relationship and share the same geometrical conditions (Wildner, 1992).

The structures are crystal-chemical isotypic if the conditions for chemical bonding are similar as well. Isomorphism is the capacity of two compounds with isotypic structures to combine into mixed crystals, or when the procedure of exchange of the atoms is able to be carried out continuously. It has been advised against using this term in this context because it is also used to describe some other phenomena (Hantsch, Winkler, & Milman, 2003). The terms "homeotypic" and "isotypic" are used to describe two structures that are similar but do not satisfy the aforementioned criteria for isotypism because of different symmetry, distinct atomic positions took by multiple kinds of atoms (substitution derivatives), or distinct geometric conditions (which change axes ratios, angles, or atomic coordinates). An example would be the substitution derivatives C (diamond), ZnS (zinc blende), and Cu_3SbS_4 (famatinite) (Hybler, Petricek, Jurek, Skala, & Cisarova, 1997).

The most effective way to determine the relationships among homeotypic structures is to make use of those relationships' symmetry.

Two ionic compounds are referred to as "antitypes" when they share the same structure type but exhibit "exchange of cations and anions," where the anions of one compound occupy the cationic positions of the other and vice versa (Benchrifa & De Pape, 1990). For instance, Li_2O crystallizes as the "anti-CaF_2 type," with the Li^+ ions taking the positions of the F ions in CaF_2 and the O^{2} ions taking those of the Ca^{2+} ions (McConnell, 1943).

1.2. CHEMICAL BONDS

In contemporary chemistry, the idea of the chemical bond is crucial. In its original, classical shape, which was painstakingly and gradually developed over the span of the 19th century, molecules were described as collections of linked atoms. Though it was developed for a considerable amount of time without any knowledge of the underlying physics, the concept proved to be incredibly helpful for interpreting, organizing, and forecasting chemical facts (Sanderson, 2012). Toward the end of the century, this "black box" situation started to alter. G. J. Stoney determined the fundamental charge of electricity in 1881 and gave it the name "electron" in 1891. W. Weber proposed that the atom is made up of positive and negative electric charges in 1894. J. J. Thomson, J. S. Townsend, and W. Wiechert, measured the electron's charge in 1897. William Thomson (Lord Kelvin) as well as J. J. Thomson developed the "plumcake" atomic model in 1902–1904. In this model, electrons are dispersed across a consistent sphere of positive electricity. In 1904, Nagaoka proposed that the positive charge is located in the atom's nucleus, with the electrons circling it (Bader, 2009).

In 1911, the planetary model of an atom was verified by E. Rutherford. An atom's valence, according to R.A. Begg's theory from 1904, is defined by the number of electrons it has gained or lost; the total has to be 8, and the most positive valence matches to the Group (column) number in the Periodic Table (Ruedenberg, 1962). J. Stark suggested that an atom's outer (or "valence") electrons decide its chemical properties in 1908, and W. Ramsay discussed the electronic structure of the bond between atoms in his essay "Electron as the Element." Last but not least, in 1913, N. Bohr proposed a theory in which the majority of the electrons in a molecule are arranged surrounding its nuclei, similar to how they are in created atoms, and only the outer electrons spin within the axes connecting atoms, creating the chemical bond (Bader, 1975). Kossel (1916) made the significant claim that there is a gradual transition from purely polar compounds (like HCl) to usually non-polar ones (like H_2) by exchanging electrons between atoms in order to finish both of their external electronic shells (Phillips, 1970).

The establishment of the chemical bond theory saw significant contributions from Linus Pauling (1916–1931); Coulson (1952); Hückel (1930–1932); Hund (1928–1931); Lewis (1916–1931); and Mulliken (1928–1931). The concepts of hybridization, the polarity and property of a bond, how much of the double-bond nature, the theory of the regional electro-neutrality of atoms, and impacting valence were all developed by Linus Pauling, who played a significant role in the development of modern structural chemistry (Silvi & Savin, 1994). Pauling's VB theory generally adopted the (implicit) belief that atoms continue to be present in molecules as easily distinguishable entities. The widely held notion that molecules are made completely of nuclei and electrons (orbitals) as opposed to actual atoms became apparent afterward, with the advancement of the theory of molecular orbitals. Remember that the energy required to split a benzene molecule into six nuclei alongside charge +6, six protons, as well as 42 electrons, all at limitless separations, is 6,07,837 kJ/mol (depending on MO calculations) (Zhang, Ying, Wu, Hiberty, & Shaik, 2009).

The energy needed to divide benzene into its six carbon as well as six hydrogen atoms, or its (experimental) atomization energy, is only 5463(3) kJ/mol, or a fraction of 1% of the previous. To compare, crystalline benzene has an intermolecular cohesion value of 44 kJ/mol as measured by the sublimation enthalpy (Kutzelnigg, 1973). As a result, atoms in molecules (AIMs) are just as "real" as those in crystals. Bader and Parr later revived the idea of AIMs, this time on the basis of contemporary quantum mechanics. Despite all of quantum chemistry's achievements, structural chemistry is still primarily an experimental field of study (Bader, 2009).

1.3. TYPES OF BONDS: COVALENT, IONIC, METALLIC

Covalent, donor-acceptor, metallic, ionic, and van der Waals are the traditional classifications of chemical bonds, but an actual bond is always made up of a few, or all, of the types. Only the most basic substances and the homonuclear bonds in symmetric molecules, which make up a very small portion of all known substances, can be found to have pure covalent bonding (Hiberty, Danovich, & Shaik, 2022). Alkali metal halides come near to having pure ionic bonds, but they do not entirely exist because covalence always exists to some extent. However, it is essential to start with the perfect types in order to comprehend real chemical bonds. This section will focus on the experimental properties of various chemical bonds and only temporarily discuss the mathematical foundations of interatomic interactions (Oleś, Pfirsch, Fulde, & Böhm, 1987) (Figure 1.4).

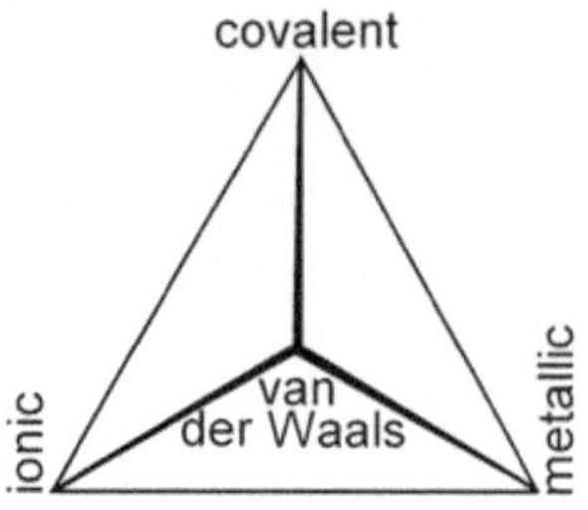

Figure 1.4. Tetrahedron of chemical bond types.

Source: Wikipedia, Creative Commons License.

1.4. IONIC BOND

The electrostatic attraction of ions with opposing charges produces the ionic bond. The crystal lattice energy U (MX), that can be measured experimentally using the Born-Haber cycle or theoretically calculated from the known net charges of ions, defines its strength in MX configuration of ionic crystals. The ionic theory elaborates many structural inorganic chemistry principles (Bader & Henneker, 1965). As a result, smaller ions (cations) fill the spaces left by larger ions (anions) in many ionic structures. Because only tetrahedral and octahedral gaps are present in this configuration, this elaborates why cations typically have a coordination number, N_c, of 4 or 6. Ionic crystals get dissolved in polar liquids (such as water) because of high solvation (hydration) heat because electrostatic interactions are strong, causing ionic crystals to have excessive fusion (melting) temperatures as well as elevated atomization energies. Low refractive indices, high broadband gaps, atomic polarizations, and insulator properties are caused by a deficiency of electrons in the inter-ionic space (Schäfer, Eisenmann, & Müller, 1973).

As mentioned, Kossel established the notion that the change from ionic to covalent substances happens gradually, with covalence rising as a result of the ions' mutual polarizing effects. Fajans (1923) and his school conceptualized this idea but ultimately were unable to produce a quantitative theory (Coll & Treagust, 2003). They polarizability of ions and calculated the polarizing action of cations (Z/r^2), therefore, can be explained simply as: "only intermediate cases that are somewhat similar to completely ionic substances exist. Because ionic radii cannot be specifically defined from interatomic distances, the parameters of ideal ions are therefore not accessible through experiments." The idea of polarization was thus limited to a qualitative level (Taber, 1997).

The basic ionic model is only a rough approximation. The theoretical oxidation states of the elements in question are first assumed to be equal to the charges of the ions. Second, the spatial distribution of the ions is only affected by their relative sizes and the search for the closest contacts among ions with opposing charges. Ions are conceptualized as hard spheres that are undoubtedly present (Reed & Schleyer, 1990). The ions can be given a more realistic effective charge, e, to increase agreement with the experiment. The charges of anions in various compounds with the same cations are related, for example, in alkali halides, as$(e^*_{MF}/e^*_{MCl})^2 = 1.245$, $(e^*_{MBr}/e^*_{MCl})^2 = 0.923$, $(e^*_{MI}/e^*_{MCl})^2 = 0.826$ (Hutskalov, Linden, & Čoric, 2023) (Figure 1.5).

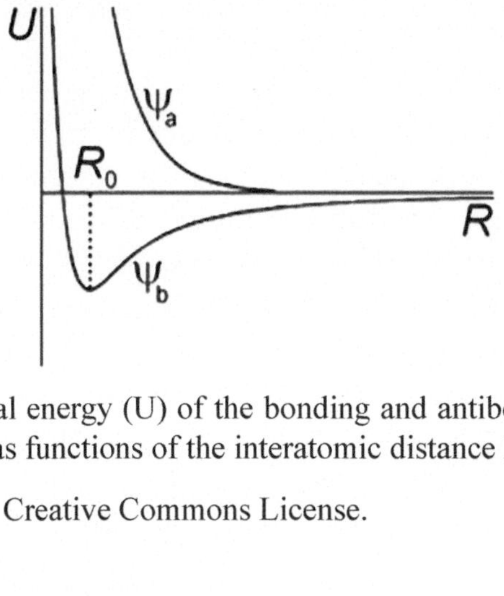

Figure 1.5. Potential energy (U) of the bonding and antibonding orbitals of a diatomic molecule as functions of the interatomic distance R.

Source: Wikipedia, Creative Commons License.

1.5. COVALENT BOND

Two electrons from usually from the same atoms (or from two non-metallic atoms), typically share electrons to form a covalent bond among two atoms. These electrons are most likely to be partially restricted in the space between the two nuclei. If these electrons have orbitals of Ψ1 and Ψ2, the bonded atoms' molecular orbitals must be their linear arrangement, with symmetric Ψb = Ψ1 + Ψ2 and antisymmetric Ψa = Ψ1 – Ψ2. This was actually illustrated by the Lewis' model, which viewed bonds as shared pairs of electrons since a maximum of two electrons can only take up an orbital (Grandbois, Beyer, Rief, Clausen-Schaumann, & Gaub, 1999). Lewis also observed that, except for H, which has two electrons in its valence shell, each stable atom typically has a maximum of eight electrons in its valence shell, including both the unshared and bonding pairs of electrons and disregarding the polarity of the bonds. Since there are just 1*s* and 3*p* orbitals in an electron shell that may take up a maximum of eight electrons between them, the octet rule has long been thought of as a law of chemistry. When a compound did not follow this rule, it was classified as either hypervalent (having more than 8 electrons) or hypovalent (having fewer than 8 electrons) (Kamaras, Itkis, Hu, Zhao, & Haddon, 2003), which is commonly one of the exceptions of the octet rule.

Contrary to Lewis' approach, the octet rule was also applied to the operational number of electrons surrounding every atom. Exceptions were justified by incorporating *d* orbitals into bond polarity, net atomic charges, and hybridization. Now, it is evident that this rule has many exceptions that don't exhibit any extraordinary properties despite being pedagogically useful. In particular, there are no obvious differences between the bonds in octet molecules and those in hypervalent molecules in terms of their fundamental makeup (Maeda, Otsuka, & Takahara, 2009). Similar to how giving net electrical charges to atoms does not do anything to help us understand the structures and characteristics of molecules or to comply with the octet rule. In a protonated amine or pyridine, the nitrogen atom contains a formal charge of +1, but the bond distances do not contract in accordance with this; actually, this atom has a slight negative charge! (Chakma & Konkolewicz, 2019).

Even though Pauling (1931); and von Antropoff (1949) predicted that potent oxidizing agents with high electronegativity could chemically awaken these, it appears that the octet rule and incorrect interpretations of quantum-mechanical concepts helped in discouraging the search for compounds of 'inert' gases for a very long time. When xenon compounds were found in 1962, this prediction was verified. Since then, hundreds of the 'rare' (previously 'inert' and afterward 'noble') gas compounds have been created, and dozens have undergone X-ray diffraction analysis. Finally, under extremely high pressure, solid xenon was changed into a metallic state (Held, Fuchs, & Studer, 2017).

Another to consider among the exception is the H_2 molecule. The H_2 molecule serves as the starting point (and occasionally the conclusion) of every modern textbook's clarification of covalent bonding because it is the most straightforward and early case studied. However, it is not an ordinary one—a fact that most students frequently fail to recognize. Since no non-bonding electrons exist in H_2, the bonding pair must also "take responsibility" for all interactions of molecules, including repulsive and attractive ones. Consequently, H_2 exhibits many peculiar properties. As a consequence, neither task is completed effectively. The order of bond significantly less than the conventional 1 is consistent with the distance between atoms, atomization energy, compressibility, and other physical properties of molecules (Green & Parkin, 2014).

Did You Know?

Ionic bonds are the strongest of all bonds. This is due to the electrostatic force that exist between them.

The HH length of bond of 0.74 Å in the H_2 molecule matches the bond order of just 0.57 Å, as determined by the distances between hydrogen and other atoms (which can make up for the absence of electron density in the bond from their own). H_2 exhibits the opposite relationships in both respects from A_2^+, which typically has lower ionization potentials than isolated atoms and higher bond dissociation energies than neutral molecules. The general VSEPR guidelines for molecular geometry are broken by hydrides of Groups 15 and 16 elements. While F-A-F angles typically widen as ligand electronegativity decreases, H-A-H angles are smaller than F-A-F angles. This impact is furthermore explained by the fact that the electron pair in the A-H bond also serves as the H ligand's non-bonding pair (Bonn & Hunger, 2021).

The noticed van der Waals radius of hydrogen is approximately 0.25 Å is less than the calculated value based on generalized radii of halogens. This demonstrates that the intermolecular repulsion is weak, while the significant departure of H_2 from Trouton's law—the enthalpy of evaporation has an unchanged ratio to the boiling point temperature—shows the concurrent weakness of intermolecular attraction. For the H_2 molecule, this ratio is up to 44.8 J $mol^{-1}K^{-1}$, linked to the average of 88 J $mol^{-1}K^{-1}$ (de Oteyza et al., 2013).

1.6. METALLIC BOND

The accessibility of electrons that move freely (previously known as valence electrons), which are shared by all atoms, is the primary characteristic of the structure of electrons of metals. Drude (around 1900) was the one who first proposed this model, considering that there were charged carriers moving between the ions having a specific velocity and colliding with each other similarly to gas molecules. He did this by applying the kinetic theory of gases to an "electron gas" in metals. It is possible to think of the metallic bond as a non-directional covalent bond. In fact, a crystal-chemical approach indicates that a change from covalent to metallic bonding may be associated with an increased coordination number, leading to a gradual delocalization of the valence electrons until they finally move from the valence into the conduction band (Anderson, Burdett, & Czech, 1994). Valence electrons of atoms should be taken out of them in order for them to freely move in the interatomic gaps of the crystal space where the metallic bond will eventually form. The prerequisite for this is $E(A\text{-}A)+I(A)<E(A^{+}...e^{-})$. A dielectric metal change happens when the A^{+} ...e^{-} interaction has more advantage as compared to the A-A bond (Grigorovich, 1989).

All of the electrons present in valence shells of atoms were thought to be free in the early theories of metals, and the metal structure was described as a lattice of cations submerged in an "electron sea." As the metallic radii are larger than those of cations, it is now known that just a portion of an atom's outer electrons is free. The metallic/core radii ratio was estimated to be 1:0.64 in a few studies of the distribution of electron density in metals. The radii of the bond of identical metals in crystalline compounds, which represent atoms with charges under ±1, are close to the operational atomic center radii in metallic structures. There is no cause to presuppose the ionization of two or more electrons from an atom because the work functions of bulk metals are always less as compared to the first ionization potentials of the related atoms (Abell, 1985).

Several research were already done in terms of high-pressure research was done on the crystal-chemical method of metallization in ionic crystals of the MX type since the late 1900s. It was determined that the distances between atoms for an inevitably condensed MX crystal are equivalent to the sum of the cationic radius of M^{+} and the regular covalent radius of X if the metallization of a material in pressure takes place when the chemical bond is damaged, it means that the compression energy gets equal to E(M-X). Because they contain a sub-lattice of neutral nonmetal atoms (X_0), metallic binary compounds are different from pure metals in this way. This implies that if the bond is covalent, M_0-X_0, as IM < IX, the M atom can be the donor of electrons in MX at high pressures. In the event that the substance M-X is ionic, then the donor must be the X-anion since AX < IM (Vahrenkamp, 1978). Experimental analysis can establish the polarity of the MX bond before pressure-induced metallization. The suggested process of metallization assumes the availability of particular structural voids that these electrons may inhabit in addition to mobile electrons since these electrons are also somewhat mobile in aromatic molecules. Considering that the structures of metals typically have high N_c values of 12, a cluster needs at least 13 metal atoms to develop metallic properties. When the number of Hg atoms attains 18, which indicates the beginning of the metallic behavior, the *s-p* band gaps in the photoelectron spectra of clusters of mercury and magnesium close (Rodriguez & Goodman, 1992).

Compound crystal structures typically have metal sub-lattices identical to or similar to those of the related pure metals. The metal lattice of the MB structure is present in MBO_4 compounds of the $CrVO_4$ structural type. As a result, the compound's structure is based on the metal skeleton, and oxygen atoms are merely added to the spaces between cations. Although there are other, more complex explanations, the minimal amount of work needed for developing the new structure on the foundation of the metal lattice can be used to explain how the derivatives inherit the structure of the parent substance (Eberhart, 1996).

The crystal-chemical method appeared to be adequate for describing the structure of the metallic bond for a very long time. The conduction band with electrons must only be partially filled, according to a physically broader approach that considers the band structures of different substances. $K_2Pt(CN)_4$ is therefore an insulator with a full band, and the Pt-Pt distance together in the chain is 3.48. On the other hand, the metallic non-stoichiometric compound $K_2Pt(CN)_4Br_{0.3}$•$3H_2O$ has a much shorter Pt-Pt distance of 2.88 Å due to the removal of electrons from the band's top, where the majority of anti-bonding interactions are found. As a result, when the band of Pt-IV is only partially filled, partial oxidation of $K_2Pt(CN)_4$ converts this compound into a metal. The insulator $K_2Pt(CN)_4Br_2$ is fully oxidized (Soler et al., 2000).

Similarly, in $La_{2x}Sr_xCuO_4$, the metal conductivity is seen at $x > 0.05$. There is yet another way for molecules to form the metal state without switching to structures that have substantial coordination numbers. Reduced molecular H_2, O_2, N_2, and halogens obtain metallic properties upon compression as a result of the shortened intermolecular distances strengthening electronic interactions. These are referred to as "molecular metals." In 3-center A-A-A orbitals and chain-like structures that form as the molecules get closer to one another, an increase in Nc from 1 (A_2) to 2 (–A–) causes a linear delocalization of the valence electrons (Gervasio, Bianchi, & Marabello, 2004).

1.7. SUMMARY

The study of chemical compounds' structures, also known as stereochemistry or structural chemistry, is a branch of chemistry that focuses primarily on molecular structures. A suitable description of the atoms' spatial distribution is required to determine the molecular makeup of a chemical compound. The coordination number is the total amount of coordinated atoms—the nearest neighbor's atoms. A crystal is a larger network of connected atoms with a periodic structure in three dimensions. The crystal structure is the arrangement of the atoms in space. The crystal lattice is created by connecting the periodically recurring atoms of one type in 3 space directions to a 3-dimensional grid. A collection of atomic coordinates, or the three coordinates x, y, and z, describe an atom's location inside a unit cell. When two compounds have identical atomic distributions and symmetry, their crystal structures are said to be isotypic. Coulson, Hund, Lewis, Hückel, Mulliken, and, particularly, Linus Pauling made significant contributions to the expansion of the chemical bond theory. Covalent, donor-acceptor, ionic, van der Waals and metallic, are the traditional classifications of chemical bonds, but a real bond is always made up of some, or even all, of these types. The Coulomb attraction of ions with opposing charges produces the ionic bond. Two electrons, one from each atom, typically combine to form a covalent bond between two atoms. The presence of freely moving electrons, which are shared by all of the atoms, is the primary characteristic of the electron structure of metals.

Activity 1.1

Define what is an isotope and give five examples of isotopes.

REVIEW QUESTIONS:

1. Give an overview of structural chemistry.
2. Explain coordination number and coordination polyhedral.
3. What is crystal structure?
4. What is isotypism?
5. Briefly explain ionic bond and metallic bond.

REFERENCES

1. Abell, G., (1985). Empirical chemical pseudopotential theory of molecular and metallic bonding. *Physical Review B, 31*(10), 6184.
2. Alvarez, S., Alemany, P., Casanova, D., Cirera, J., Llunell, M., & Avnir, D., (2005). Shape maps and polyhedral interconversion paths in transition metal chemistry. *Coordination Chemistry Reviews, 249*(17–18), 1693–1708.
3. Anderson, W. P., Burdett, J. K., & Czech, P. T., (1994). What is the metallic bond? *Journal of the American Chemical Society, 116*(19), 8808–8809.
4. Aspinall, G., (1959). Structural chemistry of the hemicelluloses *Advances in Carbohydrate Chemistry* (Vol. 14, pp. 429–468). Elsevier.
5. Bader, R. F., (1975). Molecular fragments or chemical bonds. *Accounts of Chemical Research, 8*(1), 34–40.
6. Bader, R. F., (2009). Bond paths are not chemical bonds. *The Journal of Physical Chemistry A, 113*(38), 10391–10396.
7. Bader, R., & Henneker, W., (1965). The ionic bond. *Journal of the American Chemical Society, 87*(14), 3063–3068.
8. Barger, G., & Dale, H. H., (1910). Chemical structure and sympathomimetic action of amines. *The Journal of Physiology, 41*(1–2), 19.
9. Batsanov, S. S., & Batsanov, A. S., (2012). *Introduction to Structural Chemistry*. Springer Science & Business Media *3*(5), 70–82.
10. Benchrifa, R., & De Pape, R., (1990). Isotypism of the triclinic TI8Mo10O34 and (NH4) 8Mo10O34 molybdates. *Acta Crystallographica Section C: Crystal Structure Communications, 46*(4), 728–728.
11. Birle et al., (1968). Crystal structures of natural olivines. *American Mineralogist: Journal of Earth and Planetary Materials, 53*(5–6), 807–824.
12. Blatov, V., (2000). Search for isotypism in crystal structures by means of the graph theory. *Acta Crystallographica Section A: Foundations of Crystallography, 56*(2), 178–188.
13. Bonn, M., & Hunger, J., (2021). Between a hydrogen and a covalent bond. *Science, 371*(6525), 123–124.
14. Brunner, G. T., & Schwarzenbach, D., (1971). Zur Abgrenzung

der Koordinationssphäre und Ermittlung der Koordinationszahl in Kristallstrukturen. *Zeitschrift für Kristallographie-Crystalline Materials, 133*(1–6), 127–133.

15. Carter, F. L., (1978). Quantifying the concept of coordination number. *Acta Crystallographica Section B: Structural Crystallography and Crystal Chemistry, 34*(10), 2962–2966.
16. Chakma, P., & Konkolewicz, D., (2019). Dynamic covalent bonds in polymeric materials. *Angewandte Chemie International Edition, 58*(29), 9682–9695.
17. Coll, R. K., & Treagust, D. F., (2003). Investigation of secondary school, undergraduate, and graduate learners' mental models of ionic bonding. *Journal of Research in Science Teaching: The Official Journal of the National Association for Research in Science Teaching, 40*(5), 464–486.
18. Cotton, F., & Bergman, J., (1964). Eight-coordinate complexes of cobalt(II). A principle influencing the occurrence of high coordination numbers. *Journal of the American Chemical Society, 86*(14), 2941–2942.
19. Cowan, S., et. al., (1992). Crystal structures explain functional properties of two *E. coli* porins. *Nature, 358*(6389), 727–733.
20. de Oteyza, D. G., Gorman, P., Chen, Y.-C., Wickenburg, S., Riss, A., Mowbray, D. J., & Rubio, A., (2013). Direct imaging of covalent bond structure in single-molecule chemical reactions. *Science, 340*(6139), 1434–1437.
21. Dehnicke, K., & Greiner, A., (2003). Unusual complex chemistry of rare□earth elements: Large ionic radii—small coordination numbers. *Angewandte Chemie International Edition, 42*(12), 1340–1354.
22. Dorrian, J., Newnham, R., Smith, D., & Kay, M., (1972). Crystal structure of Bi4Ti3O12. *Ferroelectrics, 3*(1), 17–27.
23. Duke, C. B., (1996). Semiconductor surface reconstruction: The structural chemistry of two-dimensional surface compounds. *Chemical Reviews, 96*(4), 1237–1260.
24. Eberhart, M., (1996). The metallic bond: Elastic properties. *Acta Materialia, 44*(6), 2495–2504.
25. Gervasio, G., Bianchi, R., & Marabello, D., (2004). About the topological classification of the metal–metal bond. *Chemical Physics Letters, 387*(4–6), 481–484.

26. Glusker, J. P., & Trueblood, K. N., (2010). *Crystal Structure Analysis: A Primer* (Vol. 14). Oxford University Press.
27. Grandbois, M., Beyer, M., Rief, M., Clausen-Schaumann, H., & Gaub, H. E., (1999). How strong is a covalent bond? *Science, 283*(5408), 1727–1730.
28. Green, M. L., & Parkin, G., (2014). Application of the covalent bond classification method for the teaching of inorganic chemistry. *Journal of Chemical Education, 91*(6), 807–816.
29. Grigorovich, V. K., (1989). *The Metallic Bond and the Structure of Metals 1*(4), 1–12.
30. Hansch, C., & Fujita, T., (1964). p-σ-π analysis. A method for the correlation of biological activity and chemical structure. *Journal of the American Chemical Society, 86*(8), 1616–1626.
31. Hantsch, U., Winkler, B., & Milman, V., (2003). The isotypism of BeH_2 and SiO_2: An ab initio study. *Chemical Physics Letters, 378*(3–4), 343–348.
32. Hartshorn, R. M., Hey-Hawkins, E., Kalio, R., & Leigh, G. J., (2007). Representation of configuration in coordination polyhedra and the extension of current methodology to coordination numbers greater than six (IUPAC technical report). *Pure and Applied Chemistry, 79*(10), 1779–1799.
33. Held, P. A., Fuchs, H., & Studer, A., (2017). Covalent□bond formation via on□surface chemistry. *Chemistry – A European Journal, 23*(25), 5874–5892.
34. Henrich et al., (2005). Proprotein convertase models based on the crystal structures of Furin and Kexin: Explanation of their specificity. *Journal of Molecular Biology, 345*(2), 211–227.
35. Hiberty, P. C., Danovich, D., & Shaik, S., (2022). A conversation on new types of chemical bonds. *Israel Journal of Chemistry, 62*(1–2), e202000114.
36. Hoppe, R., (1970). The coordination number – an "inorganic chameleon." *Angewandte Chemie International Edition in English, 9*(1), 25–34.
37. Hutskalov, I., Linden, A., & Čorić, I., (2023). Directional ionic bonds. *Journal of the American Chemical Society 1*(4), 1–12.
38. Hybler, J., Petricek, V., Jurek, K., Skala, R., & Cisarova, I., (1997). Structure determination of vistepite SnMn 4 B 2 Si 4 O 16 (OH)

2; isotypism with bustamite, revised crystallographic data and composition. *The Canadian Mineralogist, 35*(5), 1283–1292.

39. Kamaras, K., Itkis, M., Hu, H., Zhao, B., & Haddon, R., (2003). Covalent bond formation to a carbon nanotube metal. *Science, 301*(5639), 1501–1501.
40. Katritzky, A. R., Kuanar, M., Slavov, S., Hall, C. D., Karelson, M., Kahn, I., & Dobchev, D. A., (2010). Quantitative correlation of physical and chemical properties with chemical structure: Utility for prediction. *Chemical Reviews, 110*(10), 5714–5789.
41. King, R. B., (1969). Chemical applications of topology and group theory: I. Coordination polyhedra. *Journal of the American Chemical Society, 91*(26), 7211–7216.
42. King, R. B., (1970). Chemical applications of topology and group theory. IV. Polyhedra for coordination numbers 10–16. *Journal of the American Chemical Society, 92*(22), 6460–6466.
43. King, R., (1996). The shapes of coordination polyhedra. *Journal of Chemical Education, 73*(10), 993.
44. Kutzelnigg, W., (1973). The physical mechanism of the chemical bond. *Angewandte Chemie International Edition in English, 12*(7), 546–562.
45. Liebau, F., (2012). *Structural Chemistry of Silicates: Structure, Bonding, and Classification*: Springer Science & Business Media, *2*(3), 37–55.
46. Lobo, R. F., (2003). Introduction to the structural chemistry of zeolites *Handbook of Zeolite Science and Technology* (pp. 80–112): CRC Press.
47. Lockard, R. E., & Lingrel, J. B., (1969). The synthesis of mouse hemoglobin chains in a rabbit reticulocyte cell-free system programmed with mouse reticulocyte 9S RNA. *Biochemical and Biophysical Research Communications, 37*(2), 204–212.
48. Lohkamp et al., (2006). The crystal structures of dihydropyrimidinases reaffirm the close relationship between cyclic amidohydrolases and explain their substrate specificity. *Journal of Biological Chemistry, 281*(19), 13762–13776.
49. Maeda, T., Otsuka, H., & Takahara, A., (2009). Dynamic covalent polymers: Reorganizable polymers with dynamic covalent bonds. *Progress in Polymer Science, 34*(7), 581–604.
50. Marçalo, J., & De Matos, A. P., (1989). A new definition of coordination number and its use in lanthanide and actinide coordination and

organometallic chemistry. *Polyhedron, 8*(20), 2431–2437.

51. McConnell, D., (1943). Isomorphism and isotypism among silicates and phosphates. *Science, 97*(2508), 98–99.
52. McGreevy, R., Baranyai, A., & Ruff, I., (1986). A less arbitrary determination of coordination numbers in disordered systems. *Physics and Chemistry of Liquids an International Journal, 16*(1), 47–54.
53. McLachlan, A. D., (1972). A mathematical procedure for superimposing atomic coordinates of proteins. *Acta Crystallographica Section A: Crystal Physics, Diffraction, Theoretical and General Crystallography, 28*(6), 656–657.
54. McRee, D. E., (1999). XtalView/Xfit – a versatile program for manipulating atomic coordinates and electron density. *Journal of Structural Biology, 125*(2–3), 156–165.
55. Niebel, K., & Venables, J., (1974). An explanation of the crystal structure of the rare gas solids. *Proceedings of the Royal Society of London – A Mathematical and Physical Sciences, 336*(1606), 365–377.
56. O'Keeffe, M., (1979). A proposed rigorous definition of coordination number. *Acta Crystallographica Section A: Crystal Physics, Diffraction, Theoretical and General Crystallography, 35*(5), 772–775.
57. Oleś, A., Pfirsch, F., Fulde, P., & Böhm, M., (1987). Strength of electron correlations in different types of chemical bonds. *Zeitschrift für Physik B Condensed Matter, 66*, 359–362.
58. Parthé, E., (1980). Crystal-chemical formulae for simple inorganic crystal structures. *Acta Crystallographica Section B: Structural Crystallography and Crystal Chemistry, 36*(1), 1–7.
59. Pauling, L., & Corey, R. B., (1951). Atomic coordinates and structure factors for two helical configurations of polypeptide chains. *Proceedings of the National Academy of Sciences, 37*(5), 235–240.
60. Pena, M., & Fierro, J., (2001). Chemical structures and performance of perovskite oxides. *Chemical Reviews, 101*(7), 1981–2018.
61. Peresypkina, E., & Blatov, V., (2000). Molecular coordination numbers in crystal structures of organic compounds. *Acta Crystallographica Section B: Structural Science, 56*(3), 501–511.
62. Phillips, J., (1970). Ionicity of the chemical bond in crystals. *Reviews of Modern Physics, 42*(3), 317.
63. Pitzer, K., (1960). The nature of the chemical bond and the structure of

molecules and crystals: An introduction to modern structural chemistry. *Journal of the American Chemical Society, 82*(15), 4121–4121.

64. Pyykko, P., (1988). Relativistic effects in structural chemistry. *Chemical Reviews, 88*(3), 563–594.
65. Reed, A. E., & Schleyer, P. V. R., (1990). Chemical bonding in hypervalent molecules. The dominance of ionic bonding and negative hyperconjugation over d-orbital participation. *Journal of the American Chemical Society, 112*(4), 1434–1445.
66. Rodriguez, J. A., & Goodman, D. W., (1992). The nature of the metal-metal bond in bimetallic surfaces. *Science, 257*(5072), 897–903.
67. Ruedenberg, K., (1962). The physical nature of the chemical bond. *Reviews of Modern Physics, 34*(2), 326.
68. Ruiz□Martínez, A., & Alvarez, S., (2009). Stereochemistry of compounds with coordination number ten. *Chemistry – A European Journal, 15*(30), 7470–7480.
69. Sanderson, R., (2012). *Chemical Bonds and Bonds Energy* (Vol. 21). Elsevier.
70. Schäfer, H., Eisenmann, B., & Müller, W., (1973). Zintl phases: Transitions between metallic and ionic bonding. *Angewandte Chemie International Edition in English, 12*(9), 694–712.
71. Serezhkin et al., (1997). The method of intersecting spheres for determination of coordination numbers of atoms in crystal structures. *Russian Journal of Inorganic Chemistry, 42*(12), 1871–1910.
72. Silvi, B., & Savin, A., (1994). Classification of chemical bonds based on topological analysis of electron localization functions. *Nature, 371*(6499), 683–686.
73. Soler, J. M., Beltrán, M. R., Michaelian, K., Garzón, I. L., Ordejón, P., Sánchez-Portal, D., & Artacho, E., (2000). Metallic bonding and cluster structure. *Physical Review B, 61*(8), 5771.
74. Stout, J., & Reed, S. A., (1954). The crystal structure of mnf2, fef2, cof2, nif2 and znf2. *Journal of the American Chemical Society, 76*(21), 5279–5281.
75. Svergun, D., Barberato, C., & Koch, M. H., (1995). CRYSOL–a program to evaluate X-ray solution scattering of biological macromolecules from atomic coordinates. *Journal of Applied Crystallography, 28*(6), 768–773.
76. Taber, K. S., (1997). Student understanding of ionic bonding: Molecular

versus electrostatic thinking. *School Science Review, 78*(285), 85–95.

77. Tizard, H. T., (1954). *Nevil Vincent Sidgwick, 1873–1952.*
78. Vahrenkamp, H., (1978). What do we know about the metal-metal bond? *Angewandte Chemie International Edition in English, 17*(6), 379–392.
79. Vranka, R. G., & Amma, E. L., (1967). Crystal structure of trimethylaluminum. *Journal of the American Chemical Society, 89*(13), 3121–3126.
80. Wells, A. F., (2012). *Structural Inorganic Chemistry*: Oxford University Press.
81. Wildner, M., (1992). Isotypism of a selenite with a carbonate: structure of the buetschliite-type compound $K_2Co(SeO_3)_2$. *Acta Crystallographica Section C: Crystal Structure Communications, 48*(3), 410–412.
82. Zhang, L., Ying, F., Wu, W., Hiberty, P. C., & Shaik, S., (2009). Topology of electron charge density for chemical bonds from valence bond theory: A probe of bonding types. *Chemistry – A European Journal, 15*(12), 2979–2989.

CHAPTER 2

Intermolecular Forces

Contents

Unit Introduction 28

2.1. Van Der Waals Interaction 28

2.2. Interdependence of the Lengths of Van Der Waals and Covalent Bonds 35

2.3. Van Der Waals Radii 37

2.4. Interactions of Donor–Acceptor 45

2.5. Hydrogen Bond 46

2.6. Summary 50

References 51

UNIT INTRODUCTION

The donor-acceptor or van der Waals connections, or the creation of hydrogen bonds, that describe the characteristics of things and heats of evaporation of molecular substances, are what causes the mutual attraction of neutral molecules or atoms across various molecules (Stone, 2013). The reader will discover many instances of structural, thermodynamic chemical and physical, properties, as well as characteristics of products of molecular interactions, in the reviews of Bent (1968) as well as Haaland. Only a few key issues and questions related to structural chemistry and thermodynamics will be highlighted here.

2.1. VAN DER WAALS INTERACTION

Only a small percentage, or sometimes even a tiny portion, of the atomization energies, which range from numerous kilojoules per mole to tens of kilojoules per mole, are contributed by exploratory sublimation heats in molecular substances (Holstein, 2001). The vdW energies are even lower in separated vdW-molecules of the R_gM, R_gX as well as $R_gR_{g'}$ kinds, where M is a metal, X is a nonmetal and R_g is a rare gas, ranging from 0.1-2.5 kJ/mol. The root of the attractive forces among non-polar molecules was discovered at the start of 20th century. Wang (1928) made the first known demonstration of the long-range attractive energy among two hydrogen atoms in 1928. This energy differs as D_6 in which D is their separation. London then gave his "general theory of molecular forces," which contained a rough formula connecting the interaction energy to the free molecule's polarizability (Stöhr, Van Voorhis, & Tkatchenko, 2019).

$$E_{v_d w} = \frac{3}{2}\frac{I_A I_B}{I_A + I_B}\frac{\alpha_A \alpha_B}{DAB} \tag{2.1}$$

where; DAB is the intermolecular (vdW) distance, A, B is the polarizability of atoms A and B, as well as $I_{A,B}$ are their ionization potentials. London demonstrated that these forces—which he named the dispersion effect—are caused by quantum-mechanical changes in the coordinates of the electrons. Afterward, theoretical facets of the distribution (vdW) interaction were considered in reviews (Roth, Neal, & Lenhoff, 1996). These constants can be combined into C_6 to rewrite Eqn. (2.1) as:

$$kD = \left(\frac{C}{E_{vdw}}\right)^{\frac{1}{6}} \tag{2.2}$$

where; $k = (2/3)1/6 = 0.935$, The experiment yields $k = 1.05$ for molecules of the Rg•A as well as Rg•M types, where M = Zn, Cd, and Hg, indicating a 10% difference from theory. According to Slater and Kirkwood's theory,

$$C_{6=}k \frac{\alpha_1\alpha_2}{\left(\frac{\alpha_1}{N_1}\right)^{\frac{1}{2}}\left(\frac{\alpha_2}{N_2}\right)^{\frac{1}{2}}} A^6 \quad (2.3)$$

$N_{1,2}$ is the total number of interactions between electrons in atoms 1 and 2, and k is a constant (Andersson et al., 1998). The equation that follows for the vdW energy was created as a result of a straightforward method for computing $N_{1,2}$ energy

$$E_w = 0.72\frac{C_6}{D_6} \quad (2.4)$$

It provides a good fit with the experiment (E_w in meV, if D in Å) A fast and straightforward way to calculate C_6. Additionally, there are empirical approximations that closely match experimental data (Hermann, DiStasio, & Tkatchenko, 2017). Consequently, by substituting D_6 with V_2 and keeping in mind that organic molecule ionization potential differences are minimal, we get:

$$E_w \sim k\left(\frac{\alpha}{v}\right)^2 = c\left(F_{LL}\right)^2 \quad (2.5)$$

F_{LL}, the Lorentz-Lorenz function, is used here. This equation, thus, reproduces the dependence of organic compound evaporation temperatures on their refractive indices. In addition to a clear distinction in bond distances and energies, the fundamentally dissimilar nature of chemical bonding as well as vdW relationship of atoms results in a substantial change of these properties upon switching from homoatomic to heteroatomic molecules (Steiner & Desiraju, 1998). For the chemical bonds, EAB > 1/2 (EAA + EBB) and dAB 1/2 (dAA+dBB), whereas this picture is reversed for the vdW interactions:

$$D_{AB} \geq \frac{1}{2}\left(B_{AA}D_{BB}\right), E^{w}_{AB} < \frac{1}{2}\left(E^{w}_{AA}E^{w}_{BB}\right) \quad (2.6)$$

The measured length of bonds in vdW complexes deviate very little from additivity, which has only recently been recognized. Although more pronounced, the differences among the additive as well as experimental dissociation energies of vdW complexes were not noticed until 1996

(Feinberg & Sucher, 1970). It directly follows from the London equation:

$$\Delta E^{w}_{AB} = \frac{1}{2}\left(E^{w}_{AA} + E^{w}_{BB}\right) - E^{w}_{AB}\frac{3}{8}\left(I_{,}\frac{\alpha_A^2}{D_A^6} + I_B\left(\frac{\alpha_B^2}{D_B^6}\right) - 4\frac{I_A I_B}{I_A I_B}\left(\frac{\alpha_A \alpha_B}{D_{AB}^6}\right)\right) \tag{2.7}$$

Because the ionization potentials is relatively less different in numerous atoms, for atoms whose size can be compared,

$$\Delta E^{w}_{AB} \approx \frac{3}{8}\frac{I}{D^6}\left(\alpha_A^2 + \alpha_B^2 - 2\alpha_B \alpha_B\right) = c\left(\alpha_A - \alpha_B\right)^2 \tag{2.8}$$

i.e., a heteroatomic vdW bond's energy is never greater than its additive value. This is only a rough estimate, though. The approximation DAADBBDAB is likely to miscalculate the denominator in Eqn. (2.1), consequently makes less the strength because of the real distances between atoms in heteroatomic vdW molecules which constantly go beyond the additive values (Steiner & Desiraju, 1998). The analysis showed that the formula worked.

$$\Delta E_w = c\Delta\alpha^{1.2}$$

It provides better conformity with experimental data for various vdW contacts, such as rare gas dimers, rare gas-metal complexes, vdW metal-metal complexes, and complexes of molecules with rare gases as well as metals. The polarizabilities of the atoms at any distance are assumed unaffected by any vdW interactions in the formula above. The calculated polarizabilities of rare gases as well as molecular substances differ in relation to the aggregate state. The relative variations for Ar and I_2 range from 0.3% to 16.8%. The polarizability of molecules can change upon condensation, for example, by 6.6% for Br_2, by 3.2% for CF_4 or $SnBr_4$ or by 3.0% for Cl_2, or 16.8% for I_2. Because the effectual molecular volume constantly becomes less during condensation, the buildup of electron density between molecules is the only possible explanation for this increase in polarizability (Gregory, 1981).

$$\Delta = \frac{\Delta R}{R}\frac{v}{\Delta v}$$

Refraction, under normal circumstances, corresponds to volume; thus, a divergence of o from 1 typically denotes a qualitative modification of the mode of bonding. Heating or compressing gases only slightly, by 102 to 104%, alters their refractions; however, the situation is very different when compressing condensed substances. As a result, solid hydrogen compressed at P = 10-130 GPa demonstrates that o = 0.31. At P = 1.5–4.5 GPa, HCl, CO_2, and glycerin compression yields o values of 0.21, 0.22, and 0.29, respectively (Morawietz, Singraber, Dellago, & Behler, 2016). Thus, a significant portion of the change in molar volumes can be attributed to the relative change in polarizability; as a result, the power of D in Eqn. (2.1) should be distinct.

Remember

An atom's effective charge must affect its vdW radius.

The N_2H_s/ratio: here H_s is the sublimation heat of an uncommon molecular substance or a gas that is condensed (A_2, CH_4, C_6H_6) and is the dissociation energy of the equivalent dimers, has been demonstrated to be relatively near to the coordination number of the questioned atoms or molecules in the solid state. As a result, vdW energy is an additive number, because Pauling discovered when he calculated Hs of Xe. The exception to this rule is helium, which has N_2 despite crystallizing in a tightly packed structure with N_c 12 (Nir, 1977). But the decreasing strength of vdW relationships in the reduced state is closely related to helium's superfluidity (Figure 2.1).

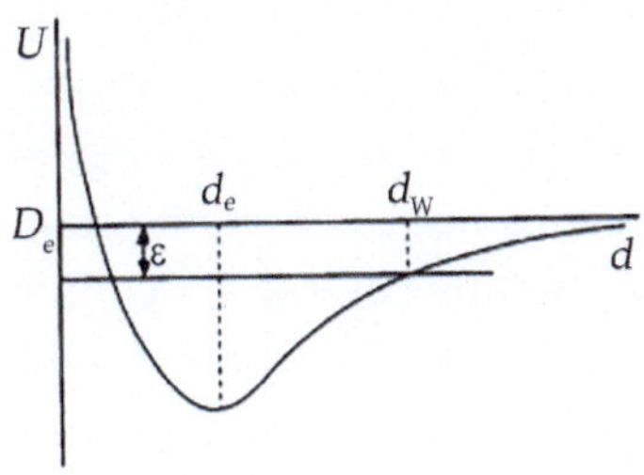

Figure 2.1. Determination of distances of Van der Waals using the Morse potential curve.

Source: Wikipedia, Creative Commons License.

The only substance that does not solidify below 0 K without external pressure is liquid helium. This is demonstrated by the substance's quantum nature, the zero-point energy (ZPE) of which is greater than the energy of the crystal lattice. Helium can be treated in a conventional manner because its macroscopic properties, such as its crystal structure as well as thermochemical traits, are not basically different from other rare gases (Gobre & Tkatchenko, 2013).

It is clear that solid helium has much longer interatomic distances as compared to the vdW molecule, in contrast to other rare gases where the dissimilarities are negligible. With liquid rare gases, the situation is comparable. Although the fact that helium expands as it solidifies has been known since 1934, the explanation for why helium atoms move farther apart during condensation needs to be further investigated (Beguin, Vernier, Chicireanu, Lahaye, & Browaeys, 2013).

Condensed helium exhibits atom oscillations with critical amplitudes comparable to those of other rare gases. As a result, the unique characteristics of helium solidification cannot be explained by the Lindemann criterion. In this situation, notice the striking resemblance between the modifications made to MX and He_2's properties as a result of the change from the molecular to the solid state: in addition to an amplification in Nc, we notice the enlargement of interatomic distances by 20–25% and a 1.5-fold increase in atomization's enthalpy (Li et al., 2019). It is known that the lowering down of density of electrons of every bond as their number rises causes the M-X bonds to lengthen while also increasing in energy. The London model of van der Waals interactions does not predict a similar outcome; in this model, the relations of atoms are solely dependent on their polarizability and size, and the number of contacts is unimportant. There exist other strategies, though (Cole, Velegol, Kim, & Lucas, 2009).

Therefore, Feynman in (1960's) noted in his research on the electrostatic clarification of chemical bonding that vdW attractive forces arise from the buildup of electron density among the atoms rather than from (London's) "interactions among oscillating dipoles"; contemporary level ab initio calculations for high level of the He_2 dimer supported his claim (Gómez-Santos, 2009). The changes in the total, kinetic, as well as potential energies as a function of the internuclear division for polar, vdW, and covalent interactions suggest a general fundamental quantum mechanism, contrary to Slater's 1930 assertion that there is a basic difference among vdW as well as covalent bonding. Later work by Bader disproved this assertion. Mulliken's "magic formula" has recently been used to calculate numerous pairs of nonmetal atoms' covalent energies (E_c) at vdW distances (I.-C. Lin, Seitsonen, Coutinho-Neto, Tavernelli, & Rothlisberger, 2009).

Slater and Kirkwood's method states that it is necessary to account for the number of electrons in isolated atoms when calculating the interaction energies among them. The fact that helium only has 2 electrons, inadequate for full-fledged interface with 12 neighbors in the crystal structure, makes it different from other rare gases. The solid state needs to be stabilized by applying external pressure after the bonds have been weakened (extended) (Vold, 1961).

For the molecules Ar_2, Kr_2, and Xe_2, Minemoto et al. (2001) calculated the transverse (αt) as well as the longitudinal (αl) axes of the polarizability ellipsoid. They discovered that the differences between these two axes—Δα, αl, αt 0.45, 0.72, and 1.23, in that order—was not consistent with the model of 2 contacting spherical atoms. Obviously,

$$¾ = \left(\frac{\alpha_1}{\alpha_2}\right)^{\frac{1}{3}} = \left(1+\frac{\Delta\alpha}{\alpha_1}\right)^{\frac{1}{3}} \leq \left(1+\frac{\Delta\alpha}{\alpha}\right)^{\frac{1}{3}} \tag{2.9}$$

When the experimental values of α and Δα and are substituted to Eqn. (2.9), the result is γ = 1.09 for all of the three components. As a result, the Xe_2 molecule, for instance, consists of 2 flattened ellipsoids instead of 2 spheres joining along the Rg-Rg vector. This suggests an interatomic interaction that is quite strong (Kannemann & Becke, 2010).

The fact that vdW complexes frequently have significant dipole moments, such as $ArSO_3$ (0.268 D), $ArCO_2$ (0.068 DN2 SO_3) (0.460 D), and, ArHCCH (0.027 D), that clearly show an electron shift from one atom to another, is also noteworthy. In fact, significant dipole moments were still observed in heteronuclear rare gas complexes like Ne-Xe (0.012 D) and Ne-Kr (0.011 D) In addition, Pyykkö have developed for the dispersion relationships in endohedral AB system, a London-type formula is needed (Henrard, Hernandez, Bernier, & Rubio, 1999). Both the static dipole polarizability of the inner system, A, as well as a novel kind of dipole polarizability, 2(B) with an r^2 radial operator, are involved. The 2nd-order energy of correction is:

$$\Delta E^{(2)} \approx -\frac{3}{4}\frac{I_A I_B}{I_A + I_B}\alpha(A)\alpha^{-2}(B) \tag{2.10}$$

The radius of B is not explicitly dependent on the new formula. The predicted interaction energies for AC60, where A He-Xe, Cd, Zn, CH_4, Hg are seen in relation to MP_2 super-molecular calculations. The clear MP_2 relationship energies for the supermolecule AB and this expression reasonably agree (Pauling & Beach, 1935). It's intriguing how these energies can be measured, but for an open-structure outer subsystem B, such as a ring- or bowl-shaped one, it is simple to do so (Sandoghdar, Sukenik, Hinds, & Haroche, 1992).

Recognizing organic solids as well as biochemical processes, including enzyme-inhibitor or enzyme-substrate interactions, requires a special focus on van der Waals interactions. The cohesion between organic molecules increases with molecular size, as do the vaporization as well as enthalpies of sublimation, boiling, and melting points. This is expected from every angle given that the number of potential atom-atom contacts as well as the number of electrons driving London's attraction rise concurrently. As a result, the standard vaporization enthalpies of common alkanes, C_nH_{2n+2}, exhibit a nearly ideal linear dependence on n, with H(vap) = 4.9 kJ mol^1 (Ambrosetti, Ferri, DiStasio, & Tkatchenko, 2016). The total amount of valence electrons per molecule correlates linearly with the overall lattice energy of organic solids, Z_v:

$$E = 1.2z_v + 20 \text{ KJ/mol}$$

2.1.1. Van der Waals in Inorganic Crystals

The interactions caused by van der Waals forces in organic crystals can be broadly divided into two categories. One theory holds that the solid stays mutually by the attraction among the atoms of neighboring molecules directly in contact; further interactions are (implicitly) considered unimportant. The alternative strategy focused on interactions that involve whole molecules or even the complete structure, are less localized and more diffused. What point on the potential curve do the nearest intermolecular distances relate to is a different but related query. The earlier perspective logically implies that they match, or at least come close to matching, the potential minimum (Liu et al., 2022). The alternative perspective, championed by Allinger (2010), contends that because the smallest contacts are shorter than de and thus repulsive, the attractive interactions between atoms located further apart are actually balanced by them. The issue is made more difficult because the exact form of the potential curve is typically unknown: only rare gases and carbon can be directly measured through experiment. The nearly identical vdW distances and energies in R_{g2} molecules and R_g crystals suggest that the vdW distance is primarily determined by the atoms in contact as well as is only marginally influenced by further nearby atoms (Lin, Huang, & Duan, 2019).

Gavezzotti & Dunitz (2009) have extensively demonstrated this crucial issue and concluded that while the immediate-contact model is adequate and essential as a 1st estimate, "intermolecular links do not exist among single atoms." As a result, they suggested switching to a more complicated estimate, the so-called *pixel* method, in place of the outdated method of finding out cohesive energy as the sum of pair-wise atom-atom interactions. Quantum-chemical techniques are used after to calculate the electron density of each individual molecule, after which the intermolecular interactions among partitions of these electron distributions, which are very small as compared to atoms, are calculated. This approach is currently more accurate and uses a lot less computer time than exhaustive quantum calculations (Dunitz & Gavezzotti, 2009). As a result, it made it possible to determine, with adequate accuracy, the equilibrium separation among 2 molecules if the rest of the crystal structure remained unchanged. It was discovered that the 1st coordination shell only contains shortened, destabilizing contacts in a small minority of instances which are reinforced by the remaining of the structure (Ganguly & Desiraju, 2008). Therefore, even though Allinger's (2010) effect exists, it is not as widespread as frequently believed. In the following sections, the effective positive charges on atoms can account for the reduction of vdW contacts in solids is illustrated. Their vdW radii will fall as well to the isolated atoms. They will eventually reach the crystallographic radii if like charges were present on the atoms of the gas-phase molecules (Batsanov, 2001) (Figure 2.2).

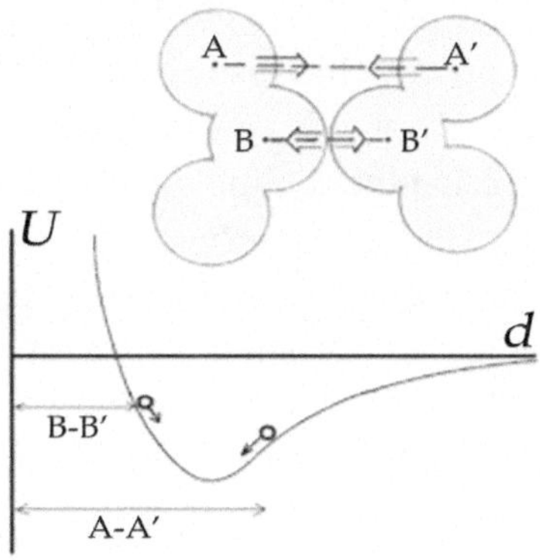

Figure 2.2. Interactions between molecules according to Allinger.

Source: Wikipedia, Creative Commons License.

2.2. INTERDEPENDENCE OF THE LENGTHS OF VAN DER WAALS AND COVALENT BONDS

It is undeniably true that organic compounds have strongly shortened distances that may not be demonstrated in the usual manner. According to experimental studies, actual bond lengths can range from the sum of covalent radii to the sum of vdW radii. Despite the irregular reduction of an I...I distance having been noticed in 1928, there is no complete theory on the matter (Jalenti, Lopez, & Azzi, 1980). The shortening of the distances was explained by a distortion of vdW spheres by some authors, including Kitaigorodskii (1965); Porai-Koshits (1980); Zefirov & Zorkii (1989). Others explained how covalent bonds between molecules caused the contraction of vdW contacts; such bonds in crystalline iodine were proven through nuclear quadrupole resonance studies. As an alternative, molecule interactions between donors and acceptors were used to explain this effect. Other studies have interpreted the reduction in intermolecular distance as proof of a novel type of interaction that has been referred to as "specific," "secondary," or the "interaction which the classical chemical bonding cannot explain" (Batsanov, Batsanov, Batsanov, & Batsanov, 2012).

In fact, these differences in bond lengths have a crystal-chemical nature and are compliant with accepted theories. Let's use the linear triatomic system I_1 I_2 I_3 as an example. Since the D vs d curve exhibits hyperbolic behavior, the O'Keeffe & Brese (1991) equation can be used to explain it.

$$-\Delta d = 0.37 \ln q \qquad (2.11)$$

where; q is the bond order, which is determined by dividing the valence by the coordination number, or v/N_c, as well as d is the change of d. We can find out from Eqn. (2.11) the q equivalent to a specific intra-molecular length of bond, and then state the intermolecular bond order as 1-q when the lengthening of the intra-molecular (short) I-I bond occurs by a partial change of electrons in the valence shell to the region between molecules. These results demonstrate that electron density transfer causes a vdW bond to change into a covalent bond (Dvorak, Ford, Suenram, Lovas, & Leopold, 1992).

When an I_2 molecule transforms into a symmetrical $I_1I_2I_3$ system, the change in d(I-I) exactly matches the reduction in q from 1 to 0.5 due to the central atom's increase in N_c from 1 to 2.

The systems X-H Y, X-Cd X, Cl-SbCl, S-S S and Br-Br Br, all displayed comparable behavior. As a result, the change of a terminal ligand into a bridging one, for example, on dimerization of $2AX_n$ A_2X_{2n}, is equivalent to the development of a regular 3-center system from a covalent as well as vdW bond, and all modifications to distances between atoms fall inside the identical range: the length of covalent bond gets higher by approximately 0.25 Å, as well as the vdW distance becomes less by approximately 1.35 Å (Batsanov, 1995). Last but not least, it has long been assumed that any interatomic distance less than ΣrvdW indicates the presence of a bonding contact. In fact, a linear correlation among Σrcov as well as ΣrvdW has even been established and utilized to determine orders of bonds (Bullett, 1975) (Figure 2.3).

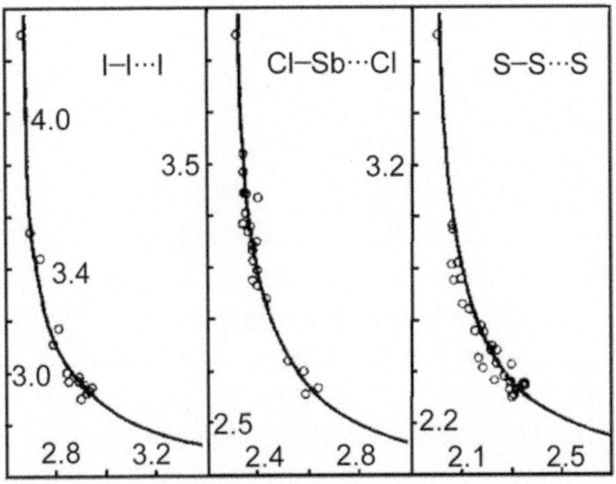

Figure 2.3. Interdependence of the lengths of Van der Waals as well as covalent bonds.

Source: Wikipedia, Creative Commons License.

2.3. VAN DER WAALS RADII

2.3.1. Introduction

The radii of van der Waals (vdW) interaction explain the separations among atoms with resolved outer electron shells or among atoms in dissimilar molecules that are not bound together. In actuality, the physics of gases is where this issue first appeared (Batsanov, 2001). As van der Waals discovered in 1879 that deviations from ideal behavior in gases (close to the critical point) can be explained by thinking of molecules are as hard spheres that are attracted to one another over larger distances, the equation of real gas follows:

$$\left(p+\frac{a}{v^2}\right)(v-b)=RT \tag{2.12}$$

In collision theory, the b is a constant related to the appropriate molecule volume.

The first step in structural chemistry was made by Magat & Mack in 1932. They gave the idea of non-valence radius (R) for that atom located at the edge of a molecule and named it the "atomic domain radius" or "Wirkungradius," suggesting that the radius finds out distances between molecules. Pauling later coined the phrase "vdW radius" to describe the intermolecular forces, subsequently called vdW interactions. In relation to this, van der Waals did not include any radii in his equation because *b* possesses the dimensionality of a volume (Vogt & Alvarez, 2014). V_o is less as compared to the same molecule's volume as determined using crystallographic vdW radii. V_o/V_m is 0.6 for organic molecules that aren't overly polar, except this ratio may be very greater for molecules that form H-bonds of higher strength (Zefirov & Zorky, 1995).

There is a ton of literature on this topic today, and there are numerous systems of vdW radii, each with a different set of sources, values, and, to some extent, physical significance. First of all, the atom lacks a distinct boundary surface. The dispersion interactions that make up the majority of the vdW forces' attractive component and Pauli exclusion that makes up their repulsive component. Only at close ranges does the repulsive force become significant, but after that it rises quite sharply (Hu, Zhou, & Robertson, 2009). The final characteristic justifies the representation of atoms in vdW relationships as hard spheres and molecules as collections of these kinds of spheres. Atoms and molecules must coexist as closely as possible in a crystal to function, but they cannot deform or penetrate one another. This is known as the Kitaigorodskii principle of close packing (Kitaigorodskii, 1965). This occurs if one molecules' bumps fit into the others' hollows (Politzer & Murray, 2021).

2.3.2. Crystallographic Van Der Waals Radii

Additionally, an atom's vdW radius was believed to be perfectly isotropic, invariant of the neighbors that have covalent bond, and solely dependent on its atomic number. Therefore, a vdW radius may be calculated by defining as half the intermolecular contact distance among comparable atoms, as well as another one can be obtained by subtracting the first from the heteroatomic distance, etc. This method was used to derive the first systems of vdW radii, that of Pauling as well as afterward Bondi (1964) as well as Kitaigorodskii (1965), from the sparse and imprecise crystallographic data available then (especially with regard to hydrogen atoms). The experimental data showed that vdW radii may differ within some tenths of an angstrom according to the structural environment (Hu, Zhou, Xie, & Robertson, 2014). It is therefore incorrect to estimate these radii based on the shortest contact distances, but it can be challenging to determine a meaningful statistical average. A histogram of every distance among specific atom types in a crystal structure can, of course, be easily assembled, except it can be a combination of 2 distinct distributions, that of (a) direct contacts as well as (b) general background. The first distribution has a bell shaped, and the second has an open end and increases monotonically with distance; as a result, the sum of the two distributions only shows a shoulder or an ill-defined maximum (Hu, Xie, & Zhou, 2010). However, the histogram's left side, which represents immediate contacts, has a point of inflection at half-height and strongly resembles a Gaussian curve. Rowland &Taylor (1996) proposed utilizing this point of inflection to describe "normal" vdW distances for a variety of pairs of atoms. They shaped a system of vdW radii to fit such distances, considering additivity. Contrary to previous ones, this one is based on a large amount of experimental data from the Cambridge Structural Database. However, while the mathematical process is straightforward, the physical significance of the radii that result is not entirely clear. However, these radii are similar to previous systems, especially Bonding's (Rowland & Taylor, 1996) (Figure 2.4).

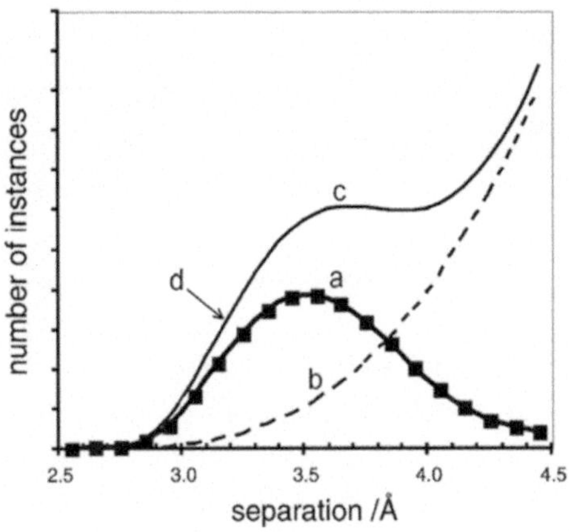

Figure 2.4. Distribution of intermolecular distances.

Source: Wikipedia, Creative Commons License.

In theory, an atom's effective charge must affect its vdW radius. In contrast, Pauling discovered that a nonmetal's vdW radius and anionic radius are identical "since the bonded atom offers an identical face to the environment present outside in ways away from its bond because the ion does in every direction" as well as utilized this rule to propose vdW radii for a few elements (Batsanov & Batsanov, 2012). Bondi (1964) used this rule, along with some other physical characteristics, in building his system of radii (he believed 0.76). He noticed as well that vdW radii go above the covalent radii (r) of the similar elements by 0.8. The equations can be used to determine this value from covalent radii:

$$r_1 - r_n = 0.3 \text{ logn}$$

$$r_1 - r_n = 0.185 \text{ logn}$$

Here; n is the order of bond and the general rule that as an intramolecular bond lengthens, the intermolecular distance decreases and, eventually, the two interactions merge. The conversion of molecules having 2 atoms into ongoing solids of the fcc or hcp kind for univalent elements results in a decrease of n from 1 to 1/12 as well as, as a result, an increase of r of 0.324 Å or 0.460 Å. Here n is the order of bond and the general rule that as an intramolecular bond lengthens, the intermolecular distance decreases and, eventually, the two interactions merge into one. The conversion of molecules with two atoms into ongoing solids of the hcp or fcc type for univalent elements results in a decrease of n from 1-1/12 as well as, as a result, an increment of r of 0.324 Å or 0.460 Å (Nyburg & Faerman, 1985).

Because there are so few (though growing) structures where metal atoms may come into close proximity with atoms of other molecules and not forming a strong bond, determining the van der Waals radii of metals is challenging (Hambley, 1998). The electronegativities of the ligands play a significant role in determining intermolecular distances, even for the same metal. As a result, the Au-Au distances in the crystals of $KAuX_4$ with X-Cl, Br, as well as I are 4.310, 4.515, and 4.843, respectively. Compared to the Pt-Pt distances in K_2PtCl_4 and K_2PtBr_4, the Pd-Pd distances in K_2PdCl_4 and K_2PdBr_4 are 4.116 and 4.309, respectively. Thus, as the ligands turn low electron-withdrawing, the effective size of the atom of metal gets higher. In addition, the M-M distance is influenced by the conformation, specifically the X–M torsion angle (Hancock, De Sousa, Walton, & Reibenspies, 2007) M–Y. Due to these issues, vdW radii for most metals remained undefined or subpar until recently. Since bond distances are more accurately defined than contact distances, inferring the vdW radii of metals from them might be useful (Hancock et al., 2007). When two spherical atoms with vdW radii overlap to form an A_2 molecule, the overlapping volume is:

$$\Delta v \frac{2}{3} \pi (R - r)^2 (2R + r) \qquad (2.13)$$

The coordination number N_c 1 is referenced here. The overlap between each pair of spheres will decrease if such molecules are combined into a close-packed metallic structure because N_c gets higher to 12 as the total amount of valence electrons stays similar. The results of experiments demonstrate that the shortening of intramolecular and intermolecular bonds that occur during the phase change from molecular solid to metal are distinct characteristics of each coordination number (Batsanov, 1999) (Figures 2.5 and 2.6).

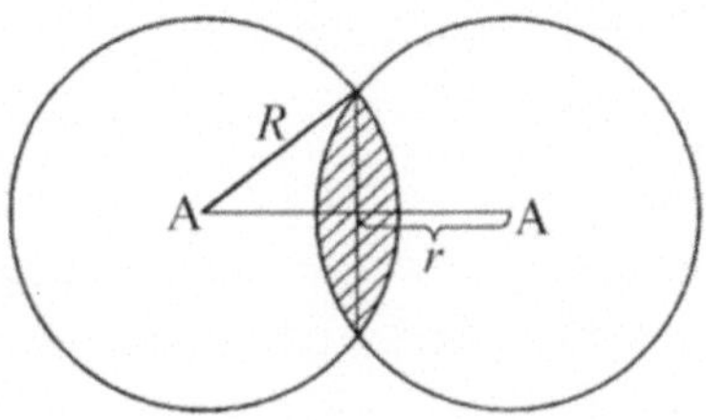

Figure 2.5. An A2 molecule shown as overlapping spheres.

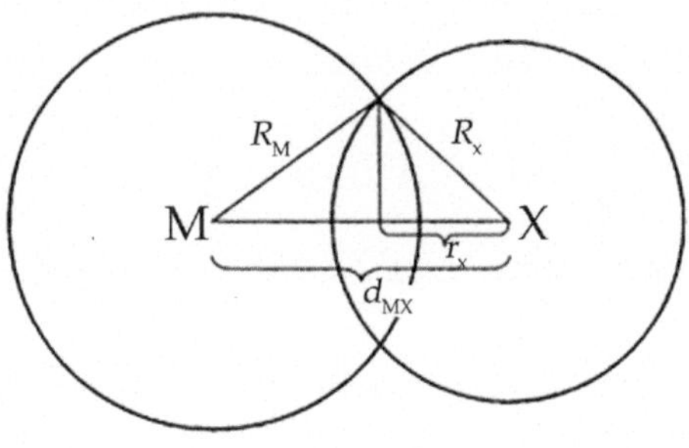

Figure 2.6. Calculation of the vdW radius of M in molecules MX.

Similar reasoning allows for estimating metal vdW radii from M-X bond distances,

$$R_M = (R_x^2 d^2 - 2\mathrm{d}r_{x)}^{\frac{1}{2}} \quad (2.14)$$

Nag et al. (1985) considered that atoms' bond valences (v) at vdW distances contain an insignificant except limited value. Hu et al. (2001) used the average volumes of the elements present in crystals covalent radii, bond valence variables, and Allinger's radii for metal-oxygen bonds to calculate the vdW radii of all elements, as well as, for the 1st time, that of the 4*f* elements. By limiting their calculations to just d1(M-O) as well as R(O) 1.40, they could study the vdW radii for all metal atoms, greatly simplifying

Nag's calculations (Taylor & Kennard, 1982). The vdW radii of elements in organic compounds are a special case. The packing factor (ρ) of molecules present in crystals can be calculated by managing atoms in molecules (AIMs) as hard spheres. According to Kitaigorodskii (1965), this value ranges from 0.65-0.77 and is strikingly same as the value of 0.7405 observed for densely packed spheres. The perfect value (0.909) for packing ideal disks was later demonstrated to be achievable by disk-shaped or large planar molecules. The packing factors ρ are obviously not to be confused with that estimated for metals as well as additional inorganic solids, which presume atoms to be spheres with much smaller metallic or covalent radii than those of the vdW (Nyburg & Faerman, 1985).

However, by considering the covalent as well as vdW contributions to ρ, we may combine the concepts of "organic" as well as "inorganic." To begin with, it can be assumed that ρ is unity (bar packing) in the plane delineated by the junction of the vdW spheres, while ρ=0.7405 exists in the other directions. The surface area of the vdW sphere is $4R^2$, and the bottom of the spherical section cut from it at a distance of r from its center is $\pi(R^2r^2)$. In the N_c sections, we have $S^* = \sigma N_c$; the ratio of the values is = $(R^2\ r^2)/4R^2$ (Kruse, Mrazikova, d'Ascenzo, Sponer, & Auffinger, 2020).

2.3.3. Equilibrium Radii of Atoms

The sum of the equilibrium radii of 2 isolated atoms, Re, corresponds to the minimum potential curve of the vdW interaction. Re differ widely because distinct interatomic potentials are employed to find out the vdW energy, and they are occasionally viewed as pure variables in molecular mechanics calculations with no severe physical significance (Kuz'mina & Struchkov, 1984). The equilibrium vdW radii, on the other hand, are physically significant, defining the surface enclosing up to 99% of the electron density of an atom or molecule, according to ab initio calculations.

$$R_{vdw} = 298\text{k} = 0.592\text{kcal} / \text{mol}$$

Because these values were chosen to work best for a limited number of compounds, there is an important distinction among the equilibrium radii of the same components utilized by various authors in their molecular mechanics computations (Wright, 2018).

2.3.4. Anisotropic Van Der Waals Radii

The intermolecular distance in crystalline iodine is dependent on the crystallographic direction, according to structural studies, and the difference is equal to 0.9. For the longitudinal (Rl) as well as transverse (Rt) vdW radii of H as well as Br, Jdanov & Zvonkova (1952) initially confirmed this result; later, this impact has been examined in several structural works (Kim, Cho, Madhav, & Kim, 2014).

Therefore, based on the structures of X_2 molecules, Nyburg & Faerman (1985) identified the anisotropic vdW radii of Halogens; from the solid-state structures of the corresponding molecules, we calculated the analogous radii for hydrogen, fluorine, nitrogen, and oxygen. It is well known that X_2 molecules may be dispersed in the crystal space in two different ways: strictly ordered orientation or isotropic distribution (through random rotation). The fcc and hcp packing of spheres with the radius R R_l 1/2 d(H-H) is how the structures of the first type are realized in $-H_1$ and $-H_2$, respectively. We obtain R_l(H) 1.516 for both instances given the cell variables of such polymorphs as well as the length of the H-H bond. There are two different types of molecules in the structures of $-F_2$ as well as $-O_2$: 25% of the molecules are chaotic, and 75% are ordered. The molecules in these structures that are ordered (parallel oriented) are spaced apart by 3.34 and 3.42, respectively. As a result, these structures' transverse vdW radii are R_T(F) 1.67 as well as R_T(O) 1.71, the distinctions among the two types of molecules (Batsanov, 2000).

From structural information, the longitudinal vdW radii of H in gas-phase Rg HX molecules were calculated. Though, in such complexes, the HX axis and Rg H direction are at an angle, so the latter separation corresponds to the middle value of R(H) (between R_l and R_t). The subsequent values of Rl(H) in Rg•HX were found out utilizing the dependence on R(H) = f (180): 1.04 for HCl, 1.04 for OH, 0.94 for HF, and 1.12 for HBr. Therefore, when the atomic charge in HX is greater, the smaller is the R_l(H). R_l(H) is equal to 0.95 in the H_2 HCl complex and 0.89 in acetylene (Eramian, Tian, Fox, Beneberu, & Kertesz, 2013).

Other gas-phase molecules' longitudinal and transverse vdW radii have also been measured experimentally. Thus, the radii of A in T-shaped vdW complexes R_gA_2 have been determined from structural data; these complexes are firm and R_t(A) are not dependent (within 0.05) on the type of R_g. Some $(A_2)_2$ dimers have side-by-side contact between the A_2 molecules, so R_t equals half of the distance among the molecular centers of mass. The R_t values are similar to the radii of A in R_gA_2 complexes, the previous radii are, on average, 0.05 larger than the latter due to different molecular packing modes, such as protuberance into hollow in R_gA_2 or protuberance toward protuberance in $(A_2)_2$ (Kim et al., 2021).

In order to account for the bend of the volumes of the central atoms in the BX_3 and MX_4 molecules as well as the intra-molecular contact radii, more complex procedures were developed for these molecules. Determining the vdW configuration of molecules that are isolated can also reveal if the anisotropy of the vdW area of atoms is intra- or intermolecular in nature. As a result, Bader et al. (2000) (discovered that the longitudinal radius (R_l) and transverse radius (R_t) of disconnected Li_2, C_2, N_2, B_2, F_2 and O_2, molecules are always smaller than each other (Strozzi, Smirnov, Pellicano, & Kovaleva, 2022) (Figure 2.7).

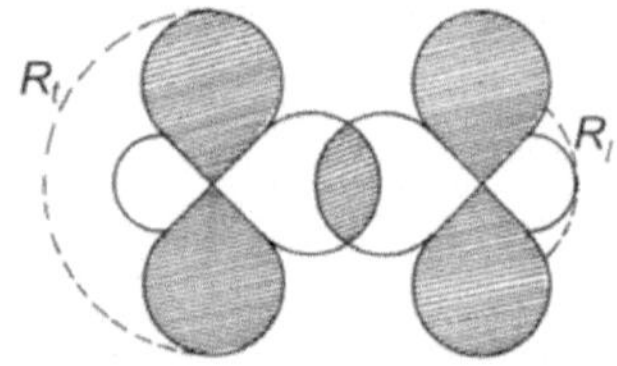

Figure 2.7. Anisotropic vdW radii.

The vdW radii are not constants; instead, they vary according to the molecular structure, chemical bond anisotropy, atomic charges, and theoretical and experimental estimates. They additionally rely on the overall state. The vdW radius of any atom depends on the effective charge, as was previously demonstrated. If the gross charge of the metal atom is 0, we possess the equilibrium vdW radius, but a negative charge on a nonmetal atom doesn't change the vdW radius much as R rX, as Pauling showed (Devi, Das, & Kumar, 2004). The vdW distance is determined only by calling atoms and relies on other atoms in the molecular or crystal space. This is rather little because the cohesive energies per "bond" in the molecular as well as crystalline positions of unusual gases are nearly the same and the distances between molecules, for this reason, are similar as well. This disproves the theory that multi-particle interactions cause the vdW contacts in solids to be less as compared to the equilibrium distances; instead, we have shown that the shortening is due to the effectual $+v_e$ charges on the atoms (Batsanov, 2002).

When such charges exist in gas-state molecules, their vdW radii decrease significantly as well and eventually converge on crystallographic radii. There may be additional factors at affect such. Due to the high strength of vdW relationship among molecules having more electronic polarizabilities, halo-gene radii decrease from CX_4 to SnX_4 as a result (Dryden et al., 2015). In liquid tetra-chlorides, a similar situation is seen: RCl is 1.75 in CCl_4, 1.63 in $SiCl_4$, and 1.53 in $GeCl_4$. Finally, because the polarizabilities of X_2 are lower as compared to those of the corresponding AX_4 molecules, the vdW radii of halogen are inferred from the structure of X_2 in the liquid state to be significantly larger than in AX_4 molecules: $R(Cl_2)$ 1.89, $R(Br_2)$ 2.03, $R(F_2)$ 1.54, $R(I_2)$ 2.23. The firmly lessened distances in some organic condensed phases, however, may be explained by such things (Bandyopadhyay & Sadhukhan, 2022).

2.4. INTERACTIONS OF DONOR–ACCEPTOR

An atom having an open orbital (A, typically a metal) and an atom having a single electron pair (D, typically a non-metal) form a donor-acceptor (D-A) bond. Coordination compounds are created as a result of DA interactions. The enigma of DA bonding is that, despite the formal electron transfer going from D to Aχ (D) > χ(A) in the reverse direction. The reason for this is that D atoms possess closed electron shells, making them vdW atoms with extremely low electronegativity ranges of χ as 0.2–1. Therefore, it is not surprising that lone electron pairs transfer to polyvalent metals with $\chi \geq 1.5$. DA bonds have lengths that are comparable to those of regular valent bonds but have much lower energies (Gutmann & Gutmann, 1978).

This pronounced difference can be explained in this way. Molecules need to be brought close together to overcome the vdW repulsion in order to form a chemical bond. This barrier for HCl and NH_3 is 94.4 kJ/mol, which is comparable to the dissimilarity among ΔHf of iso-structural salts of NH_3 and Rb (100 kJ/mol). The Madelung electrostatic forces can cause a change of vdW complex in the solid state, such as NH_3-HX to NH_4X, compensating for this energy expenditure. Mulliken (1931) concluded that NH_3 and HCl can only interact to form a vdW complex in the gas phase, which is held mutually through forces of dispersion, and that the change to ionic NH_4Cl may only be caused by electrostatic factors (Fowkes, 1987). According to ab initio calculations, $NH_x(Me)_3$-xHX complexes are not ionic pairs because, despite H-X's slight elongation, it is always less as compared to N-H. The H-N-H distances for NH_3-HF were determined by similar calculations to be 1.01 and 1.70 Å, respectively (Emmett, Prentice, & Pantoş, 2013). Leopold et al. (1997) reviewed the alterations in DA interactions and "partially formed bonds" during the change from gas to solid state and concluded that the local (condensed) environment has the ability to significantly alter the geometry of these molecules from their gas phase geometry if the distorting effect makes the molecular dipole moment higher. This reduces the overall energy in polar or polarizable media, offsetting the energy cost of the deformation (Beaujuge, Amb, & Reynolds, 2010).

As mentioned, the energy of dissociation of DA complexes ranges from 1 to 200 kJ/mol is chracaterized to be in a very wide range. However, these values are not the energies of the M-X bonds; rather, they are the Hf of 2 molecules, much like the heat required to form NaCl from the molecules Na_2 as well as Cl_2. The DA process of bonding 2 molecules in a complicated structure is identical to that of the formation of dimer molecules (Bent, 1968).

2.5. HYDROGEN BOND

As a final thought, consider the hydrogen bond. There have been a lot of books and articles written about the attachment of polar molecules via an H atom. As Pauling said, under certain circumstances, "a hydrogen atom is drawn by quite strong forces to 2 atoms, in place of just 1, so it can be thought of to be acting as a bond among them. What we refer to is the hydrogen bond. The term "H-bond" is now utilized much more broadly. It is frequently employed to explain phenomena that are neither primarily due to nor specifically related to H-bonding. Therefore, it is necessary to define the terminology first (Alkorta, Rozas, & Elguero, 1998).

Which characteristics of substances are frequently connected to hydrogen bonds? The melting point lowers down with the molecular mass in the order H_2Te, H_2S, and H_2 Se (50, 65, and 84°C), but it rises to 0°C for H_2O. Similar patterns can be seen in the boiling points (Tb) of these substances. The series HBr, HI, HF, and HCl are comparable in both respects. However, in these conditions, two concurrent variables coexist: the molecules' decreasing polarizability, which additionally impacts the vdW interaction, and their increasing polarity that results to increasing electrostatic of coulombic interaction (Pimentel & McClellan, 1971).

H-bonds that are undeniably present in ice and liquid water, only completely collapses at extremely elevated pressures or temperatures. Since such bonds stop molecules from being packed more densely, H_2O is one of the few substances whose density decreases at freezing. It is essential to define the energies of hydrogen bonds to gauge their impact on the structures as well chemical or physical characteristics of substances. According to several literatures, 15.5, 24, 18, and 9.6 kJ/mol are the best and accurate experimental values for the heats of dissociation of the gaseous dimers (H_2O), (HF), $(HCl)_2$ and (NH_3). Hydrogen bond energies are 2 demands of magnitude lower as compared to the chemical bond energies of H-N, H-O, H-Cl, and H-F. Therefore, it should not significantly affect changes in physical properties caused by forming H-bonds (Brown, 1976).

The data of neutron diffraction (ND) on various chemical compounds with X-H A hydrogen bonds, however, demonstrate obvious correlations among shrinking H-A distances as well as lengthening covalent X-H bonds. Analysis of structural information from in situ high-pressure ND examines reveals that the O-H as well as H-O distances exhibit similar correlations as those found for various compounds under ambient conditions. The development of the double-welled hydrogen bond potential into a single-well interest is another pressure-related effect. The symmetry of hydrogen bond can be indicated by the bulk modulus having a discontinuation at this point, which is determined by ab initio calculations and denotes that the symmetry of hydrogen bond is a 2nd-order phase transition (Perrin & Nielson, 1997).

By contrasting M-OH and M-F bond lengths in isostructural MF_n as well as $M(OH)_n$ compounds, the true contribution of H-bonds to these changes was examined. The strength of H-bonds appears to have no bearing on the difference between d(M-OH) and d(M-F) 0.08 and 0.02, respectively. This constancy of d results from the fact that the influence of H-bonds on interatomic distances is negligible when the coordination numbers, polarities of atoms, and changes in valences, are same (Weiss, Brandl, Sühnel, Pal, & Hilgenfeld, 2001).

The dissolution of ionic compounds in water has an effect on H-bonds. For instance, the interactions among water as well as ions disturb the long-range order of the H-bonding network in the bulk liquid by fragmenting or developing hydrogen bonds. The 1st solvation sphere of merely hydrated ions has more hydrogen bonds that are broken than water in liquid state, while the 1st coordination shell composed of highly hydrated ions has less. IR spectroscopy can detect H-bonds by changing the absorption band, such as that of v (OH), but this impact is in fact reliant on both the mass of linked vibrating atoms and the general structure of the molecule. The best approach for studying hydrogen bonds is to examine a substance's electronic structure because a molecule's mechanical characteristics have a negligible impact (Grabowski, 2006).

Did You Know?

The vdW radii are not constants.

One of the H^+ ion's most significant characteristics is its minuscule size, which gives rise to N_c one or two. This causes H-bonds to be saturated and directed. Despite having a low energy, the H-bond can change properties of structures that are already at or near the thermodynamic stability threshold, such as interfering with ice's tightly packed molecules, lowering symmetry, or altering the composition of substance. Thus, crystalline hydrates of ammonium salts contain less water than their K analogues due to the creation of H-bonds among hydrogen atoms in NH_4^+ ions as well as anions: NH_4F *vs* KF_4•H_2O, $(NH_4)_2CO_3$ *vs* K_2CO_3•$6H_2O$, $(NH_4)_3PO_4$•$3H_2O$ *v s*K_3PO_4•$7H_2O$, $(NH_4)_3AsO_4$ *vs* K3AsO_4•$7H_2O$, $(NH_4)_3Fe(CN)_6$•$1.5H_2O$ *vs* $K_4Fe(CN)_6$•$3H_2O$, etc. Similarly, substitution of hydrogen atoms with K in $[H_nAO_m]^{x-}$ anions decreases the number of water molecules of crystallization (Kuhn, 1952).

$K_3AsO_4 10H_2O$ $K_3PO_4 7H_2O K_2CO_3$•$6H_2O$ KF•$4H_2O$ K_2HAsO_4•H2O K_2HPO_4•$3H_2O$ $KHCO_3$ KHF $KH_2AsO_4 KH_2PO_4$

Conversely, the number of molecules of water in hydrates rises having the n-v_e charge on the oxygen atom in the AO_n-anion.

KNO_3, $KClO_4$, $K_2SO_4 < K_2CO_3 \cdot 6H_2O < K_3PO_4 \cdot 7H_2O < K_3AsO_4 \cdot 10H_2O$

However, take note that the H_nAO_m acids become stronger in reverse order due to the weakening effect of H-bonds on proton dissociation (Jeffrey & Saenger, 2012).

Hydrogen bonds may create with additional atoms possessing $\chi 2$, especially with metals that may function as both donors as well as acceptors in D-H M as well as M-H A systems. Up until this point, we have only taken in to account hydrogen bonds with highly electronegative acceptors, like F, O, or N. Thousands of records from the CSD have been used to thoroughly examine the geometries of interactions of D-H X-M (D C, N, and O; X halogen), as well as the analogs in which metal halide groups (X-M) are substituted by halocarbon groups (X-C) or halide ions (X) (Hobza & Havlas, 2000). The relative power of the D-H•• •X-M hydrogen bonds has been determined in the order listed below:

D–H···F–MD–H···Cl–M≥D–H···Br–M>D–H···I–M,

Which displays an impressively strong qualitative arrangement with a pattern in intramolecular N-H X-Ir hydrogen bond energies (X halogen). The

D-H H-M interactions (D N, O), also referred to as "di-hydrogen bonds" or "proton-hydride bonds," are extremely fascinating. If the hydridic hydrogen atom bonds to an electropositive primary group element (such as B, Al, or Ga), they include an interaction among protic (δ+) as well as hydridic (δ–) hydrogen atoms (Steiner & Desiraju, 1998) (Figure 2.8).

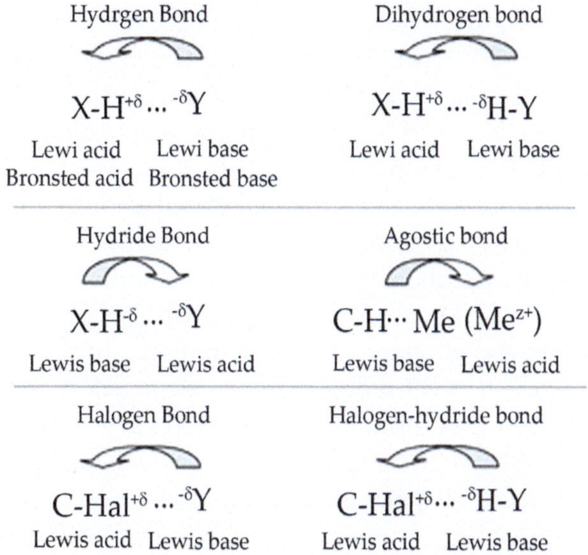

Figure 2.8. Relations of the H-atom with atoms of other elements.

Grabowski (2006) concluded that there are two ways to explain the hydrogen-bonding interaction. One approach is to view it primarily as an electrostatic interaction. The alternative viewpoint acknowledges electrostatic interaction but maintains that other types of interaction energy terms, primarily transfer of charge and dispersive energies, are also significant. Though, as was already mentioned, the wide range of structures of compounds having hydrogen bonds may be boiled down to a few straightforward laws, according to which bond lengths vary depending on the changing coordination number. Similar to hydrogen, halogen atoms frequently interact with other atoms that have lone electron pairs. This outcome, known as a "halogen bond" by similarity with an H-bond, is thought to be crucial for the self-assembly of compounds containing halogens (Scheiner, 1997).

Activity 2.1

Make a table of comparison of different molecular interactions.

2.6. SUMMARY

The formation of hydrogen bonds or the van der Waals or donor-acceptor interactions, which define the characteristics of things and heats of evaporation of molecular substances, are what cause the mutual attraction of neutral molecules or atoms across various molecules. According to experimental studies, real bond lengths can range from the sum of covalent radii to the sum of vdW radii. In fact, these differences in length of bonds are crystal-chemical in nature and are compliant with accepted theories. Van der Waals (vdW) radio explains the separations among atoms with resolved outer electron shells or among atoms in different molecules that are not bound together. The first step in structural chemistry was made by Magat & Mack in 1932. They gave the idea of non-valence radius (R) for that atom which is located at the edge of a molecule and called it the "atomic domain radius" or "Wirkungradius." The sum of the equilibrium radii of two isolated atoms, Re, corresponds to the minimum of the potential curve of the vdW interaction. vdW radius may be calculated by defining as half the intermolecular contact distance among comparable atoms, and another one can be obtained by subtracting the first from the heteroatomic distance, etc. The intermolecular distance in crystalline iodine is dependent on the crystallographic direction. An atom having an open orbital as well as an atom with a single electron pair form a donor-acceptor (D-A) bond.

Review Questions:

1. Give a brief description of Van der Waals interaction.
2. Write a detailed note on crystallographic Van der Waals radii.
3. What is meant by equilibrium radii of atoms?
4. What are anisotropic Van der Waals radii?
5. What is hydrogen bonding?

REFERENCES

1. Alkorta, I., Rozas, I., & Elguero, J., (1998). Non-conventional hydrogen bonds. *Chemical Society Reviews, 27*(2), 163–170.
2. Ambrosetti, A., Ferri, N., DiStasio Jr, R. A., & Tkatchenko, A., (2016). Wavelike charge density fluctuations and van der Waals interactions at the nanoscale. *Science, 351*(6278), 1171–1176.
3. Andersson, Y., Hult, E., Rydberg, H., Apell, P., Lundqvist, B. I., & Langreth, D. C., (1998). Van der Waals interactions in density functional theory. *Electronic Density Functional Theory: Recent Progress and New Directions*, 243–260.
4. Bandyopadhyay, P., & Sadhukhan, M., (2022). Modeling coarse-grained van der Waals interactions using dipole-coupled anisotropic quantum Drude oscillators. *Journal of Computational Chemistry*.
5. Batsanov, S. S., (2000). Anisotropy of atomic Van der Waals radii in the gas-phase and condensed molecules\sp {1}. *Structural Chemistry, 11*(2–3), 177–183.
6. Batsanov, S. S., (2001). Van der Waals radii of elements. *Inorganic Materials, 37*(9), 871–885.
7. Batsanov, S. S., Batsanov, A. S., Batsanov, S. S., & Batsanov, A. S., (2012). Intermolecular forces. *Introduction to Structural Chemistry*, 227–274.
8. Batsanov, S., (1995). Van der Waals radii of elements from the data of structural inorganic chemistry. *Russian Chemical Bulletin, 44*, 18–23.
9. Batsanov, S., (1999). Calculation of van der Waals radii of atoms from bond distances. *Journal of Molecular Structure: THEOCHEM, 468*(1–2), 151–159.
10. Batsanov, S., (2002). Correlation between the anisotropy of the electronic polarizability of molecules and the anisotropy of the Van Der Waals atomic radii. *Russian Journal of General Chemistry, 72*(8), 1153–1156.
11. Beaujuge, P. M., Amb, C. M., & Reynolds, J. R., (2010). Spectral engineering in π-conjugated polymers with intramolecular donor–acceptor interactions. *Accounts of Chemical Research, 43*(11), 1396–1407.
12. Beguin, L., Vernier, A., Chicireanu, R., Lahaye, T., & Browaeys, A., (2013). Direct measurement of the van der Waals interaction between two Rydberg atoms. *Physical Review Letters, 110*(26), 263201.

13. Bent, H. A., (1968). Structural chemistry of donor-acceptor interactions. *Chemical Reviews, 68*(5), 587–648.
14. Bondi, A. v., (1964). van der Waals volumes and radii. *The Journal of Physical Chemistry, 68*(3), 441–451.
15. Brown, I., (1976). On the geometry of O–H–O hydrogen bonds. *Acta Crystallographica Section A: Crystal Physics, Diffraction, Theoretical and General Crystallography, 32*(1), 24–31.
16. Bullett, D., (1975). Chemical pseudopotential approach to covalent bonding. ii. Bond lengths and bond energies in diamond, silicon and graphite. *Journal of Physics C: Solid State Physics, 8*(17), 2707.
17. Cole, M. W., Velegol, D., Kim, H.-Y., & Lucas, A. A., (2009). Nanoscale van der Waals interactions. *Molecular Simulation, 35*(10–11), 849–866.
18. Devi, T. G., Das, A., & Kumar, K., (2004). Raman anisotropic bandwidth study of C–O stretching vibration of methyl isobutyl ketone: role of van der Waals' volume of the interacting systems. *Spectrochimica Acta Part A: Molecular and Biomolecular Spectroscopy, 60*(1–2), 211–216.
19. Dryden, D. M., Hopkins, J. C., Denoyer, L. K., Poudel, L., Steinmetz, N. F., Ching, W.-Y., & French, R. H., (2015). van der Waals interactions on the mesoscale: Open-science implementation, anisotropy, retardation, and solvent effects. *Langmuir, 31*(37), 10145–10153.
20. Dunitz, J. D., & Gavezzotti, A., (2009). How molecules stick together in organic crystals: Weak intermolecular interactions. *Chemical Society Reviews, 38*(9), 2622–2633.
21. Dvorak, M., Ford, R., Suenram, R., Lovas, F. J., & Leopold, K. R., (1992). van der Waals vs. covalent bonding: microwave characterization of a structurally intermediate case. *Journal of the American Chemical Society, 114*(1), 108–115.
22. Emmett, L., Prentice, G. M., & Pantoş, G. D., (2013). Donor–acceptor interactions in chemistry. *Annual Reports Section" B" (Organic Chemistry), 109*, 217–234.
23. Eramian, H., Tian, Y.-H., Fox, Z., Beneberu, H. Z., & Kertesz, M., (2013). On the anisotropy of van der Waals atomic radii of O, S, Se, F, Cl, Br, and I. *The Journal of Physical Chemistry A, 117*(51), 14184–14190.
24. Feinberg, G., & Sucher, J., (1970). General theory of the van der Waals

interaction: A model-independent approach. *Physical Review A, 2*(6), 2395.

25. Fowkes, F. M., (1987). Role of acid-base interfacial bonding in adhesion. *Journal of Adhesion Science and Technology, 1*(1), 7–27.
26. Ganguly, P., & Desiraju, G. R., (2008). Van der Waals and polar intermolecular contact distances: Quantifying supramolecular synthons. *Chemistry – An Asian Journal, 3*(5), 868–880.
27. Gobre, V. V., & Tkatchenko, A., (2013). Scaling laws for van der Waals interactions in nanostructured materials. *Nature Communications, 4*(1), 2341.
28. Gómez-Santos, G., (2009). Thermal van der Waals interaction between graphene layers. *Physical Review B, 80*(24), 245424.
29. Grabowski, S. J., (2006). *Hydrogen Bonding: New Insights* (Vol. 3): Springer.
30. Gregory, J., (1981). Approximate expressions for retarded van der Waals interaction. *Journal of Colloid and Interface Science, 83*(1), 138–145.
31. Gutmann, V., & Gutmann, V., (1978). *The Donor-Acceptor Approach to Molecular Interactions* (Vol. 228): Springer.
32. Hambley, T. W., (1998). van der Waals Radii of Pt(II) and Pd(II) in molecular mechanics models and an analysis of their relevance to the description of axial M••• H (– C), M••• H (– N), M••• S, and M••• M (M= Pd(II) or Pt(II)) interactions. *Inorganic Chemistry, 37*(15), 3767–3774.
33. Hancock, R. D., De Sousa, A. S., Walton, G. B., & Reibenspies, J. H., (2007). Metal-ion selectivity produced by C-Alkyl substituents on the bridges of chelating ligands: The importance of short H–H nonbonded Van Der Waals contacts in controlling metal-ion selectivity. A thermodynamic, molecular mechanics, and crystallographic study. *Inorganic Chemistry, 46*(11), 4749–4757.
34. Henrard, L., Hernandez, E., Bernier, P., & Rubio, A., (1999). van der Waals interaction in nanotube bundles: consequences on vibrational modes. *Physical Review B, 60*(12), R8521.
35. Hermann, J., DiStasio Jr, R. A., & Tkatchenko, A., (2017). First-principles models for van der Waals interactions in molecules and materials: Concepts, theory, and applications. *Chemical Reviews, 117*(6), 4714–4758.

36. Hobza, P., & Havlas, Z., (2000). Blue-shifting hydrogen bonds. *Chemical Reviews, 100*(11), 4253–4264.
37. Holstein, B. R., (2001). The van der Waals interaction. *American Journal of Physics, 69*(4), 441–449.
38. Hu, S.-Z., Xie, Z.-X., & Zhou, Z.-H., (2010). 70 years of crystallographic van der Waals Radii. *Acta Physico-Chimica Sinica, 26*(7), 1795–1800.
39. Hu, S.-Z., Zhou, Z.-H., & Robertson, B., (2009). Consistent approaches to van der Waals radii for the metallic elements. *Zeitschrift für Kristallographie, 224*(8), 375–383.
40. Hu, S.-Z., Zhou, Z.-H., Xie, Z.-X., & Robertson, B. E., (2014). A comparative study of crystallographic van der Waals radii. *Zeitschrift für Kristallographie-Crystalline Materials, 229*(7), 517–523.
41. Jalenti, R., Lopez, A., & Azzi, A., (1980). Van der Waals' volumes from densities and vaporization enthalpies. *Thermochimica Acta, 35*(3), 315–324.
42. Jeffrey, G. A., & Saenger, W., (2012). *Hydrogen Bonding in Biological Structures*. Springer Science & Business Media *1*(3), 15–24.
43. Kannemann, F. O., & Becke, A. D., (2010). van der Waals interactions in density-functional theory: Intermolecular complexes. *Journal of Chemical Theory and Computation, 6*(4), 1081–1088.
44. Kim, H., Cho, W. J., Madhav, M. V., & Kim, K. S., (2014). Anisotropic charge distribution and anisotropic van der Waals radius leading to intriguing anisotropic noncovalent interactions. *Scientific Reports, 4*(1), 1–8.
45. Kim, S. E., Mujid, F., Rai, A., Eriksson, F., Suh, J., Poddar, P., & Zhong, Y., (2021). Extremely anisotropic van der Waals thermal conductors. *Nature, 597*(7878), 660–665.
46. Kitaigorodskii, A. I., (1965). The principle of close packing and the condition of thermodynamic stability of organic crystals. *Acta Crystallographica, 18*(4), 585–590.
47. Kruse, H., Mrazikova, K., d'Ascenzo, L., Sponer, J., & Auffinger, P., (2020). Short but weak: The Z-DNA lone-pair··· π conundrum challenges standard carbon Van der Waals Radii. *Angewandte Chemie, 132*(38), 16696–16703.
48. Kuhn, L. P., (1952). The hydrogen bond. I. Intra-and intermolecular hydrogen bonds in alcohols. *Journal of the American Chemical Society, 74*(10), 2492–2499.

49. Kuz'mina, L. G., & Struchkov, Y. T., (1984). Structural chemistry of organomercury compounds. Role of secondary interactions. *Croatica Chemica Acta, 57*(4), 701–724.

50. Leopold, K. R., Canagaratna, M., & Phillips, J. A., (1997). Partially bonded molecules from the solid state to the stratosphere. *Accounts of Chemical Research, 30*(2), 57–64.

51. Li, B., Yin, J., Liu, X., Wu, H., Li, J., Li, X., & Guo, W., (2019). Probing van der Waals interactions at two-dimensional heterointerfaces. *Nature Nanotechnology, 14*(6), 567–572.

52. Lin, I.-C., Seitsonen, A. P., Coutinho-Neto, M. D., Tavernelli, I., & Rothlisberger, U., (2009). Importance of van der Waals interactions in liquid water. *The Journal of Physical Chemistry B, 113*(4), 1127–1131.

53. Lin, Z., Huang, Y., & Duan, X., (2019). Van der Waals thin-film electronics. *Nature Electronics, 2*(9), 378–388.

54. Liu, L., Gong, P., Liu, K., Nie, A., Liu, Z., Yang, S., & Huang, L., (2022). Scalable van der Waals encapsulation by inorganic molecular crystals. *Advanced Materials, 34*(7), 2106041.

55. Minemoto, T., Matsui, T., Takakura, H., Hamakawa, Y., Negami, T., Hashimoto, Y., & Kitagawa, M., (2001). Theoretical analysis of the effect of conduction band offset of window/CIS layers on performance of CIS solar cells using device simulation. *Solar Energy Materials and Solar Cells, 67*(1–4), 83–88.

56. Morawietz, T., Singraber, A., Dellago, C., & Behler, J., (2016). How van der Waals interactions determine the unique properties of water. *Proceedings of the National Academy of Sciences, 113*(30), 8368–8373.

57. Nir, S., (1977). Van der Waals interactions between surfaces of biological interest. *Progress in Surface Science, 8*(1), 1–58.

58. Nyburg, S., & Faerman, C., (1985). A revision of van der Waals atomic radii for molecular crystals: N, O, F, S, Cl, Se, Br and I bonded to carbon. *Acta Crystallographica Section B: Structural Science, 41*(4), 274–279.

59. Pauling, L., & Beach, J., (1935). The van der Waals interaction of hydrogen atoms. *Physical Review, 47*(9), 686.

60. Perrin, C. L., & Nielson, J. B., (1997). "Strong" hydrogen bonds in chemistry and biology. *Annual Review of Physical Chemistry, 48*(1), 511–544.

61. Pimentel, G. C., & McClellan, A., (1971). Hydrogen bonding. *Annual*

Review of Physical Chemistry, 22(1), 347–385.

62. Politzer, P., & Murray, J. S., (2021). The use and misuse of van der Waals radii. *Structural Chemistry, 32*, 623–629.
63. Porai-Koshits, M. A., (1980). Crystal chemistry and stereochemistry of monobasic carboxylates of the transition metals. *Journal of Structural Chemistry, 21*(3), 369–401.
64. Roth, C. M., Neal, B. L., & Lenhoff, A. M., (1996). Van der Waals interactions involving proteins. *Biophysical Journal, 70*(2), 977–987.
65. Rowland, R. S., & Taylor, R., (1996). Intermolecular nonbonded contact distances in organic crystal structures: Comparison with distances expected from van der Waals radii. *The Journal of Physical Chemistry, 100*(18), 7384–7391.
66. Sandoghdar, V., Sukenik, C., Hinds, E., & Haroche, S., (1992). Direct measurement of the van der Waals interaction between an atom and its images in a micron-sized cavity. *Physical Review Letters, 68*(23), 3432.
67. Scheiner, S., (1997). *Hydrogen Bonding: A Theoretical Perspective*: Oxford University Press on Demand.
68. Steiner, T., & Desiraju, G. R., (1998). Distinction between the weak hydrogen bond and the van der Waals interaction. *Chemical Communications*, *8*, 891–892.
69. Stöhr, M., Van Voorhis, T., & Tkatchenko, A., (2019). Theory and practice of modeling van der Waals interactions in electronic-structure calculations. *Chemical Society Reviews, 48*(15), 4118–4154.
70. Strozzi, M., Smirnov, V. V., Pellicano, F., & Kovaleva, M., (2022). Nonlocal anisotropic elastic shell model for vibrations of double-walled carbon nanotubes under nonlinear van der Waals interaction forces. *International Journal of Non-Linear Mechanics, 146*, 104172.
71. Taylor, R., & Kennard, O., (1982). Crystallographic evidence for the existence of CH. cntdot. cntdot. cntdot. O, CH. cntdot. cntdot. cntdot. N and CH. cntdot. cntdot. cntdot. Cl hydrogen bonds. *Journal of the American Chemical Society, 104*(19), 5063–5070.
72. Vogt, J. r., & Alvarez, S., (2014). van der Waals radii of noble gases. *Inorganic Chemistry, 53*(17), 9260–9266.
73. Vold, M. J., (1961). The effect of adsorption on the van der Waals interaction of spherical colloidal particles. *Journal of Colloid Science, 16*(1), 1–12.

74. Wang, S. C., (1928). The problem of the normal hydrogen molecule in the new quantum mechanics. *Physical Review, 31*(4), 579.
75. Weiss, M. S., Brandl, M., Sühnel, J., Pal, D., & Hilgenfeld, R., (2001). More hydrogen bonds for the (structural) biologist. *Trends in Biochemical Sciences, 26*(9), 521–523.
76. Wright, A. C., (2018). The structural chemistry of B_2O_3. *Physics and Chemistry of Glasses-European Journal of Glass Science and Technology Part B, 59*(2), 65–87.
77. Zefirov, Y. V., & Zorkii, P. M., (1989). Van der Waals radii and their application in chemistry. *Russian Chemical Reviews, 58*(5), 421.
78. Zefirov, Y. V., & Zorky, P. M., (1995). New applications of van der Waals radii in chemistry. *Russian Chemical Reviews, 64*(5), 415.
79. Zhdanov, G. S., & Zvonkova, Z. V., (1952). Crystal structures of the higher oxides of metals of the first group of the periodic system. In: *Dokl. Akad. Nauk SSSR* (Vol. 82, pp. 743–746).

CHAPTER 3

Polymorphism and Phase Transition

Contents

3.1. Thermodynamic Stability 60
3.2. Kinetic Stability 60
3.3. Polymorphism 61
3.4. Energies of Phase Transitions 68
3.5. Melting Heats of Compounds 69
3.6. Sublimation Heats of Elements and Compounds 70
3.7. Evaporation Heats of Compounds 73
3.8. Enthalpies of Phase Transformations 75
3.9. Summary 76
Review Questions: 77
References 78

3.1. THERMODYNAMIC STABILITY

A structure is thermodynamically inert if the free enthalpy of the reaction ΔG for changing its structure into any other framework is positive. A structural arrangement can only be stable within a definite range of pressure and temperature because ΔG depends upon the transition enthalpy ΔH and the transition entropy ΔS. A phase change will eventually occur when ΔG becomes negative in relation to another structure due to temperature and/or pressure changes (Schellman, 1987). This could be a phase change from one modification of a solid to a different solid or a change in the aggregate state.

Thermodynamic relationships suggest:

$$\ddot{A}G = \Delta H\, T\Delta S \; and \; \Delta H = \Delta Up + \Delta V \qquad (3.1)$$

The following rules can explain the dependence of thermodynamically stable structures on temperature and pressure:

Low-order T structures will be more preferred as the temperature rises. As a result of the positive transition entropy ΔS that is involved in their formation, the term T ΔS has a significant impact on the value of ΔG. For instance, two modifications of hexahalides, such as that in MoF_6, are known as in their solid state; one has atomic particles having clearly defined atomic arrangements while the other have molecular rotation around their centers of gravitational field (Selbach, Einarsrud, & Grande, 2009). The later modification is thermodynamically more stable at higher temperatures because the order is lesser. The order is lower in the liquid state and lowest in the gaseous public. Thus, increasing the temperature will cause the substance to reach the melting phase leading to evaporation (Grimme, Huenerbein, & Ehrlich, 2011).

Structures with a lower volume, or those with a higher density, benefit from higher pressures *p*. ΔH will become negative because the process of their establishment contains a reduction in the volume (negative ΔV). For illustration, diamond particles with a density of 3.51g cm^{-3} is more stable at very high pressures than graphite (density=2.26 g cm^{-3}) (Mauzerall, 1960).

3.2. KINETIC STABILITY

When the rate of change from one structure to another is negligible, a thermodynamically unstable framework may still exist. In this instance, we refer to the structure as kinetically stable, inert, or metastable (Schmidt et al., 2017). According to the Arrhenius equation, the rate constant *k* depends on the temperature T as well as the activation energy E_a.

$$K = K_0 e^{E_a/RT}$$

When a significant ratio $E_{a/}RT$ leads to a low k, we have a kinetic stable system. The kinetic stabilization of any structure is possible at sufficiently low temperatures. Kinetic stability is arbitrary and negligible.

Glasses are typically metastable materials. They exhibit macroscopic form stability, just like crystalline solids, but due to their structure and some of their physical characteristics, they must be viewed as very viscous liquids. Only by making extensive atomic movements can they change to a thermodynamically more stable structure, while cross-linking severely restricts atom mobility (Castro, 2013).

Only because many substances are metastable which, therefore, can be studied under normal circumstances, as thermodynamically unstable structures and properties t (Hubbard & Schlom, 1996).

3.3. POLYMORPHISM

Isomers are molecules with a comparable composition but different structures while Polymorphism is the term used to describe the corresponding phenomenon in crystalline solids. Therefore, polymorphism occurs when substances tend to crystallize into various crystalline forms and structures. Thus, it is usually deduced that polymorphism is a type of isomerism, while the main difference lies in the fact that polymorphs are two or more crystals with the same chemical composition while isomers are organic molecules of the same molecular composition but differ in structure or spatial configuration. The chemical as well as physical characteristics of the polymorphs differ along with their arrangement of atoms in space. The structural variations can range from minor adjustments to the position of molecules to a totally dissimilar atomic configuration (Ford, 1945).

Roman numerals, such as tin-I, tin-II, etc., or lower case Greek letters, such as α, β..., e.g., α-sulfur, β-sulfur, are frequently used to denote different modifications of a compound. Numerous polymorphic minerals have been given meaningless names, such as the SiO_2 forms of tridymite, cristobalite, β quartz, coesite, stishovite and keatite (Purohit & Venugopalan, 2009).

The use of "*pearson*" symbols, which are more systematic but not always clear, is advised through the IUPAC (International Union of Pure and Applied Chemistry). The lower-case letter for the crystal system, the upper case letter for the kind of lattice centering, as well as the number of atoms in the unit cell make up a *pearson* symbol. For instance, the orthorhombic, face-centered molecule of sulfur (α-sulfur) has 128 positions in each unit cell (Bernstein, 2011).

Polytypes are polymorphic kinds with structures that have various stacking sequences of similar layers. The preparation and crystallization conditions determine which polymorphic shape of a compound forms. Some of the influencing factors are the synthesis method, temperature, pressure, solvent type, cooling or heating rate, fusion or gas phase, crystallization from solution, and the presence of seed crystals (Brittain, 2012).

The *Ostwald* step rule states that when a compound with multiple possible crystallizations crystallizes, it may initially take the form of a thermodynamically unstable modification under the circumstances before changing to the more stable form. For instance, selenium undergoes a chemical change in a solution that forms elemental selenium, which produces a red precipitate that is made up of Se8 molecules. This red polymorph then gradually transforms into the steady, gray shape polymeric chain molecules. Another example is KNO_3, which is unstable at room temperature but stable over 128°C. At room temperature, β-KNO_3 first crystallizes from a water solution before changing to α-KNO_3 after a short period of time or in response to a very small mechanical stress (Ernst, 2001).

The first alteration is, on the other hand, is dictated by the nucleation energy, which is affected by the surface energy. Nucleation energy typically decreases as surface energy does. First, to crystallize is the modification with the lowest nucleation energy. The order of crystallization of polymorphic kinds may be affected by the existence of foreign matter because the surface energy is delicately dependent on the adsorption of unrelated particles (Clegg, 2019).

The incidence of polymorphism, or the existence of many crystal alterations of a single substance, was 1st demonstrated by Mitscherlich in 1823. In this study, polymorphism coexisted with isomorphism (see below). It remained thus, determined that the MCO_3 compounds for example, Mn, Mg, Fe, Co, Zn and Cd devour the calcite structure with compound $N_c(M) = 6$ while those with M = Sr, Pb, and Ba, or for cations with greater extents, take on the aragonite structure with $N_c(M) = 9$. Morphotropy is the even alterations in the structure produced by modifications in the chemical configuration of a homologous sequence in mixtures (Vasiliev & Laptash, 2012). Both known modifications of $CaCO_3$ gave these structural classes their names: calcite, a mineral, low-temperature, and aragonite, a rock crystal and high-temperature form. A common example of polymorphism is the transformation of calcite into aragonite. In addition, aragonite is a good example of kinetic factor significance because, although it is thermodynamically unstable relative to calcite under ambient conditions, the transition is kinetically hindered and takes place over a geological time period (Boulineau et al., 2010).

In the structural point of view, polymorphic variations will be divided further into two categories: "reconstructive transitions," in which the atoms' nuclei change, and as "displacive shifts," in which positions of atoms are the only one that are slightly altered while the nuclei stays the same. The former transformations are crucial for structural chemistry. The designs of the thermodynamic stability of the stages are very challenging because the energies of phase transitions only account for a few per cent of the atomization energy of the solid particles, at most. Crystal-chemical methods are currently the most successful in estimating the causes and effects of polymorphism (Gholivand, Afshar, Shariatinia, & Zare, 2010).

The thermodynamic 'order' of a phase change that is found out by the Gibbs free energy derivative order is another method to categorize phase transitions. In this book, first-order phase transitions—pressure-induced transformations involving a discontinuous change in volume—are the main topics of interest. The additional byproduct of the Gibbs free energy with respect to the volume, that is inversely proportional to the compression, is unpredictable in second-order phase changes. It is often very challenging to tell the difference between a weakly 1st order transition and a phase transition of second order experimentally (Gerasimova et al., 2020) (Figures 3.1 and 3.2).

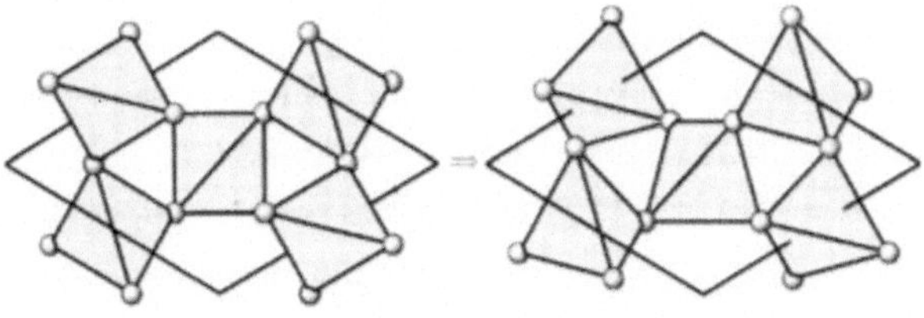

Figure 3.1. An example of reconstructive phase transition.

Source: Wikipedia, Creative Commons License.

Given the structural configurations of anionic crystal as a local packing of anions along with cations in the spaces, it is easy to wrap up the ratios of both cation as well as anion radii ($k_r = r_+/r_-$) that resembles to a variety of coordination polyhedra. Therefore, for a triangle the lesser boundary is $kr = 2\sqrt{3} - 1 = 0.155$; for a tetrahedron, $kr = \sqrt{6}/2 - 1 = 0.255$; for an octahedron, kr = $\sqrt{2}-1$ = 0.414; for a cube, kr = $\sqrt{3}-1$= 0.732. A phase shifts occurs with gradual increase in the Nc and packing density when an ionic compound undergoes compression, basically a compression of anions. When a smaller cation is swapped out for a larger one (in the case of morphotropy), a similar phenomenon takes place. Because chemical bonding is never always in pure ionic form while the oxygen ion is rigid enough; a simple model like this does not always accurately foresee the organization of atomic particles in the crystal structures, even for the molecules of oxides (Hayashi, Ozawa, & Hamaguchi, 2003).

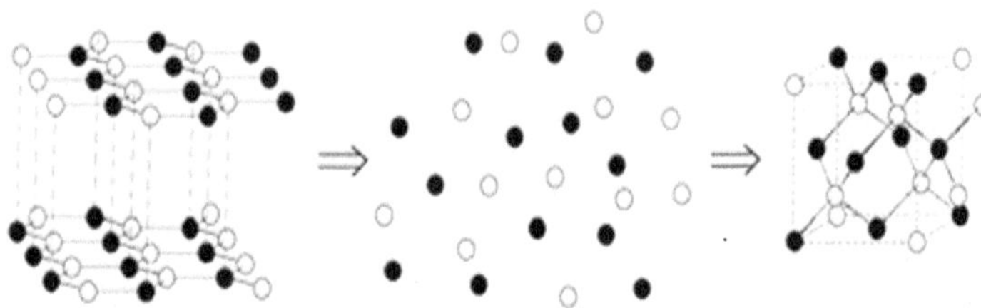

h-BN Atomiziation c-BN

Figure 3.2. An example of displacive transition.

The following picture appears when coordination numbers are taken into account in halide structures: N_c(MF) is typically lesser than N_c(MX) value where X is the halogen such as Cl, I or Br; however, $N_c(MF_n)$ is more smaller than $N_c(MX_n)$ even though the ionic theory forecasts the following succession (F being smaller): N_c(MCl) > Nc(MBr) > N_c(MF) > N_c(MI). The anion-anion repulsion, a structural destabilizing factor that depends on the effective charges of atom (e) as well as the anion-anion distances, causes the ionic model to diverge from reality (Belenkov & Greshnyakov, 2014). When defining the Coulombic (electrostatic) energy, it is mandatory to correct for modifications in the $(e^*)^2$ in order to connect more pronounced impact of attraction just to the ionic size. Important ratios of a variety of $(e)^2$ value for alike cations in alkaline halides are counting, e.g., value of F/Cl is 1.245, I/Cl have 0.826, and value of Br/Cl is 0.923 (Zelaya, Stevens, & Dowd, 2010). These effective values known as the energetic radii of anions are formed through by of the perfect ionic radii by these variables:

F^-	Cl^-	Br^-	I^-	O_2^-	S_2^-	Se_2^-	Te_2^-
2.30	2.25	2.20	2.15	3.05	2.35	2.25	2.15Å

It is clear from this situation that an anions' most potent ionic radii are in a different order of magnitude from those of their crystal-chemical counterparts. Therefore, N_c is less prevalent in MF compared to in other MX halides, which makes sense (Vogel & Nilges, 2021).

The geometrical estimates of N_c may be further enhanced by utilizing the actual radii of atoms in the polar molecules, moreover hardness, as well as considering the effect of repulsion. Born-Landé theory states that the main factor $f_n = n/(n - 1)$, here n is the Born coefficient of repulsion that determines the hardness of an ion. These factors are equivalent to 1.25 for He-type ions, 1.167 for Ne-type ions, 1.091 specific for Xe^- and Au^+ ions, 1.125 for Ar^- and Cu^+ ions while 1.111 specific for Kr and Ag^+ ions. By knowing the bond ionization values which allows one to compute the actual ionic radii, multiply those results by the hardness factor, and then determine the actual kr (Khanmohammadi, Garmarudi, Moazzen, & Ghasemi, 2010). For about 80% of structures, the coordination numbers thus determined match the ones that were actually observed. These substances have the $N_c <$ 4 in their structures, whereas in ionic structures which have $N_c > 4$ due to the lack of dense atoms packing caused by a produced covalent bond (or its significant contribution) (Bernstein, 1993). To distinguish between structures with $N_c > 4$ and 6, Phillips and Lukowsky who first utilized the ionicity factor $f_i = 0.785$ as a criterion. Considering that Wemple's equation connects the ionicity value of Phillips' and the bond ionicity calculated by Pauling (i):

$$i^2 = \frac{f_i}{(\varepsilon - 1)} \quad (3.2)$$

Phillips proved that the threshold of ionicity factor is equal to 0.5, where ε symbol exhibits dielectric constant (crystalline compounds have ε whose value is equal to 4 for with polar bonds). The ionicity value of the substances must first be estimated in order to determine the calculation method to use. If ionicity factor exceeds 0.5, then the geometrical criterion (constant=k_r) should be used; otherwise, the structure will have $N_c < 4$. With this method, the expected and experimental coordination are in perfect agreement (Malakhova, Sukhikh, Rakhmanova, & Vinogradova, 2022).

The crystal-chemical method is unable to forecast the parameters of polymorphic modifications due to difficulties in accurately calculating bond ionicity and atom radii in polar bonds, on top of a comprehensive physical theory that is still lacking (Groom & Allen, 2014).

As a result, structural diagrams with diverse coordinates, like electronegativity symbolized as (χ), atomic radius symbolized as (r), and the amount of valence electrons, etc., were developed using statistical methods. While Pettifor (1989) described the arrangements of A_nB_m compounds using the concept of "chemical electronegativities" A and B, Burdett et al. (1987) used the Coulombic (C) as well as homo-polar (E_h) elements of the atomization energy as coordinates *n* (Cruz-Cabeza & Bernstein, 2014). The *n* no. of coordinates, in which *n* is the real mean basic quantum number, allows for the observation of various atomic structure forms, such as 2D maps. The coordinates Z/r_i and χ have been proposed for complex compounds. Sproul (1994) identified the locations of covalent, metal, and ionic, substances by plotting binary compounds with the coordinates and on a map. Burdett considered a few electronic features of the structural maps. Tosi (2002) provided the structural maps for the various compounds for example, melted halides which are polyvalent molecules and also for alkaline metals. Last but not least, Villars (2003) employed 3 dimensional (3D) maps with the various coordinates such as ΔrAB, ΔχAB, and Σve, where rA or rB showing the orbitals of atomic radii like as A and B, and Σve is the sum of total number of valence electrons in such atoms (Munshi, Venugopala, Jayashree, & Guru Row, 2004).

Phase changes in organic solids typically involve shifting the orientation of molecules within the crystal space and take place at milder thermodynamic condition changes. The geometrical features of molecules don't change much, and neither carry any coordination numbers. Thus, if the first and ultimate molecules have identical structures, the growth of an additional stage can be epitaxial; otherwise, it can be disorganized (random) if the system's initial and final states differ structurally. In addition to terminology concerns, the crystal-chemical as well as thermodynamic components of polymorphism in organic compounds are taken into consideration in (Zargar, Asghari, & Dashti, 2015).

Amorphous solids are capable of changing their atomic structure in narrow range as like in alike way to crystalline solids. These so-called polymorphism cases have, thus, far been identified at pressures other than ambient. As a result, it was discovered that the N_c in liquid silicon (α-Si) and amorphous silicon (α-Si) exceeds monotonically with pressure leading to polymorphous changes were noted at later stage. Further, pairing distribution functions exhibits the primary attributes at 2.36 (the Si-Si bond) and 3.83 that is also the Si-Si distance next to the silicon nearest neighbor, which is a sign of the regional tetrahedral coordination geometry between the pressures of 3 and 13.5 GPa. The latter peak shifting to a slightly greater separation reach at P > 10 GPa as well as the earlier peak shifting to a slightly shorter separation reach at 3.40, respectively, indicate the formation of a high-coordinate Si species (Hückel, 1950).

If decompression occurs below 9.5 GPa, the change is the opposite. The period at which tetrahedral coordinated polymorphous compounds are formed, in which strong connected atoms take up space areas in the amorphous complex, may be linked to the phase transformation, according to analysis of the MD simulation configurations. $N_c(B)=3$ is present in the B_2O_3 glass structure under ambient conditions but grows with pressure and temperature. In contrast to the previous studies, solid-state NMR suggested that there are just 5% at 2 GPa and 27% at 6 GPa, X-ray diffraction data indicates that the arrangement consists of 15% of B atoms (Boron) with 30% of $N_c=4$ at 8 GPa with 300 K, and 45% at 9.5 GPa with 300 K, and 9.5 GPa at 650 K (Alvarenga et al., 2019).

Some substances exhibit crystalline polymorphism and polymorphism in combination. Furthermore, an insulator crystal phase of molecules such as I of SnI_4 shifts into a metallic phase II and reaches over 7.2 GPa, is pressure-amorphized at 15 GPa, and re-crystallizes at 61 GPa into a non-molecular, polymeric phase III with a fcc-lattice. The Ge-I distance lengthening, which starts to happen upon compression at P=20 GPa, occurs prior to the decomposition of GeI_4. Under compression, SnO_2 exhibits strong anisotropy: the a-axis is largely incompressible having a linear stiffness co-efficient of 306 GPa, while the c-axis has a co-efficient of 43 GPa. Even though SnO_2 is metastable at room temperature and breaks down into elements such as Sn and SnO_2 at maximum temperatures, the putrefaction rate under high pressure is faster than at atmospheric ambient temperatures (Repická et al., 2010).

The phase transition frequently occurs at a lower pressure when a constant pressure and a cut off deformation of the sample are present. A shock-wave compression, which consistently results in shear deformations, may also cause the phase transition at lower pressure. The partial amorphization of the sample brought on by the shear can be used to explain these effects (Tarnai, 2002). The work to destroy the initial order (W_{des}) and the work to create an alternate atomic arrangement (W_{for}) are the two parts of the phase transition's enthalpy ($\Delta H_{tr}=P_{tr}\Delta V_{tr}$):

$$\Delta H_{tr} = W_{des} + W_{for} \tag{3.3}$$

For a reconstructive stage transitions for example $W_{des}c\Delta H_m$, c is the degree of destruction of the solid and ΔH_m corresponds to atomic order of the long-range being completely destroyed (due to melting). Because the melting objects has a small-range order of 10-30Å, while automated as well as shock-wave pounding of solids decreases the particle size to no more than 100Å, leading to c values of 0.2 and 0.1 (Ghouili, Brahmia, & Ben Hassen, 2015). This value of c as well as Eqn. (3.3) indicate that a decrease in the phase change pressure is necessary.

$$\Delta P_{tr} = \mathrm{P} - W_{for} / + W_{for}\Delta V_{tr} \approx 0.2\Delta H_m / \Delta V_{tr} \qquad (3.4)$$

In the formula below, V_{tr} stands for the volume difference among the initial as well as final phases, and H_m for melting heat. As shown by Eqn. (3.4), the P_{tr} for ionic solids are, for instance, ZnO=4.1, ZnS=2.9, ZnSe=1.1 GPa, CdS=1.5, CdSe=1.4 GPa, NaF=2.5, NaCl=3.1 GPa, KF=1.9, KCl=1.2, KBr=1.1, and KI=1.3 GPa (Ludwig & Nilges, 2018).

Did You Know?

A structure may only be stable within a specific variety of pressure and temperature.

These projections show how compression with shear can lower a solid's phase transition pressure. When there is a melt, c = 1 and the impact is much greater (Arancibia, Naspi, Pucci, Rodriguez, & Di Salvo, 2020).

3.4. ENERGIES OF PHASE TRANSITIONS

The energy released during the transformation of a solid to a liquid, a fluid to a gas, as well as a solid into a gas is known as the melting heat (Hm), evaporation heat (Hv), and sublimation heat (Hs), respectively; thus, in an ideal situation ΔHs, ΔHs=ΔHv+ΔHm. In reality $\Delta H_s \approx \Delta H_v + \Delta H_m \Delta$, as these phase shifts are investigated at various temperatures. In crystalline solids, H_m reacts to the absence of long-range array from a structural perspective, whereas H_v and H_s may represent entirely distinct processes and be of various orders of magnitude (Fultz, 2020). Metals possess very high H_s, equal to the atomization energy E_a; therefore, $H_s \approx H_v > H_m$. Evaporation of metals entails the change of metallic bonds, so metals possess a very high Hs (Landau, 1936). The same situation is accurate for non-metals like diamonds having polymeric covalent structures. Because vdW relationships instead of valence bonds must be broken for solids with the molecule's structure to evaporate,

H_v is much lower and is comparable to H_m $\Delta H_s \approx \Delta H_v > \Delta H_m$ (Fazio & Van Der Zant, 2001).

3.5. MELTING HEATS OF COMPOUNDS

Cyclobutane has a ΔH_m = (kJ/mol) of 5.4, cyclopentane of 5.8, cyclohexane of 3.6, cycloheptane of 3.8, cyclooctane of 3.1, ferrocene of 17.8, and nickelocene of 19.0.

One can make an educated guess as to how much small-range order there is in liquids by knowing the melting energy (Dworkin & Bredig, 1968). Assume that solid ZnS, CsCl, and NaCl, are divided into decreasingly sized cubic blocks, and then determine how many atoms and bonds are present in each block at each stage (Gschneidner, 1969). We use crystals with these structural types as our study subjects. The mean coordination numbers (N_c) for various grains are proportions of the overall amount of bonds to the overall number of atoms in the blocks, while the ratio ΔNc/Nc indicates how many bonds in the coordination polyhedron are broken during this "grinding." These ratios are 0.09, 0.115, and 0.12 for the B3, B1, as well as B3 types of MX halides, which correspond to N_c =3.6, 5.3, and 7. The measured N_c* in the related melts is fairly close to these N_c values (Pappa, Voutsas, Magoulas, & Tassios, 2005) (Table 3.1).

Table 3.1. Melting Heats (KJ/mol) of Inorganic Compounds

V_M	M	F	Cl	Br	I	O	S
I	Li	27.1	19.8	17.7	14.6	35.6	–
	Na	33.4	28.2	26.2	23.7	47.7	19.0
	K	27.2	26.3	25.5	24.0	27.0	16.2
	Rb	25.8	24.4	23.3	22.1	21.0	–
	Cs	21.7	20.4	23.6	25.6	20.0	–
	Cu	–	7.1	7.2	7.9	65.6	9.6
	Ag	16.7	13.0	9.2	9.4	15.5	7.9
	Tl	14.0	15.6	16.4	14.7	30.3	23.0
II	Cu	55.0	20.4	–	–	49	–
	Be	4.8	8.7	18.0	21.0	86	–
	Mg	58.7	43.1	39.3	26	77	63
	Ca	30.0	28.0	29.1	41.8	80	70
	Sr	29.7	16.2	10.5	19.7	81	63
	Ba	23.4	15.8	32.2	26.5	46	63

	Zn	40	10.3	15.7	17	70	30
	Cd	22.6	48.6	33.4	15.3	–	43
	Hg	23.0	19.4	17.9	15.6	–	26
	Sn	10.5	14.5	18.0	18.0	27.7	31.6
	Pb	14.7[c]	21.8[c]	16.4[c]	23.4[c]	25.6	49.4
	Cr	34	45	45	46	–	25.5
	Mn	30	30.7	33.5	41.8	43.9	26.1
	Fe	50	43	43	39	24	31.5
	Co	58.1	46.0	43	35	50	30
	Ni	69	78	56	48	50.7	30.1
MX[d]	TiN	VN	CrN	MnN	FeN	CoN	NiN
	50	21	25	36	13.5	45	28
MX	ZnSe	ZnTe	CdTe	HgSe	HgTe	SnTe	PbTe
M_2O_3	B_2O_3	Al_2O_3	Ga_2O_3	In_2O_3	Tl_2O_3	Sc_2O_3	Y_2O_3
	24.5g	111g	99.8g	105g	105g	127	105
	La_2O_3	Ti_2O_3	As_2O_3	Sb_2O_3	Bi2O3	V_2O_3	Cr_2O_3
	125	92	30.1	61.5	28.4	140	125
M_2S_3	B_2S_3 48.5	Al_2S_3 66	Sb_2S_3 65.3	Bi_2S_3 37.2	Mo_2S_3 130	–	–
MO_2	SiO_2	GeO2	SnO2	TiO_2	ZrO2	HfO2	–
	9.6g	17.2g	23.4g	68	90	96	–
	MoO_2	WO_2	TcO_2	ReO_2	PtO_2	ThO_2	UO_2
	66.9	48.1	75.3	50.2	19.2	90	78

3.6. SUBLIMATION HEATS OF ELEMENTS AND COMPOUNDS

Table 3.2 provides a list of the sublimation (ΔH_s) heats of various elements under typical thermodynamic circumstances. H_s fluctuates on a regular basis, just like atomization energies. The ratio of H_m/H_s for metals is essentially constant, while it is roughly three times higher for semiconductor elements with ongoing covalent-bond networks and close to 1 for molecular structures (Politzer, Murray, Edward Grice, Desalvo, & Miller, 1997). The $\Delta H_m/\Delta H_s$ ratios of crystalline inorganic compounds with ongoing bond networks are comparable to those of metals. Crystalline compounds have atoms whose coordination numbers are greater than their valences, which leads to lower electron densities per bond as well as high polarity of bond. This explains why H_s depends on atomic charges, as first demonstrated by Fajans and then later by Urusov (1995). By comparing the sublimation temperatures of polar compounds to those of the related elements, where the bonds are precisely covalent, one can get rid of this dependence (Politzer, Ma, Lane, & Concha, 2005). Alkali halides serve as the best example, where a change in the crystal molecule only alters the polarity for the chemical bond. As a result, a compound's sublimation heat is greater than the sum of its constituent heats (Carson, Stranks, & Wilmshurst, 1958).

Due to inadequate vdW interactions, molecular solids have much lower sublimation heats. As the molecular size decreases, they typically increase. For elements, the sublimation heat increases with increasing polarizability, with values for N_2, Ar, and I_2 being 6.88, 7.73, and 62.44 kJ/mol, respectively. Accordingly, ion-dipole relationships or hydrogen bonds enable polar molecules to exhibit much higher H_s despite having lower polarizability (Greenwood, Perkins, & Twentyman, 1967) (Figure 3.3 and Table 3.2).

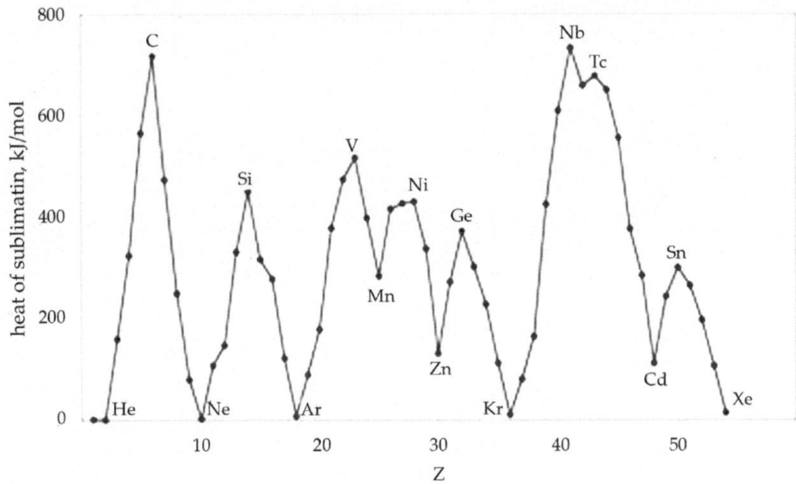

Figure 3.3. Sublimation heats of elementary solids, periods 1–5.

Table 3.2. Sublimation Heats of Metal Oxides (kJ/mol)

		Na	**K**	**Rb**	**Cs**	**Tl**
	425	385	304	272	187	198[a]
MO	**Cu**	**Be**	**Mg**	**Ca**	**Sr**	**Ba**
	462	742[a]	660[a]	679[a]	589[a]	420[a]
	Zn	Cd	Hg	Sn	Pb	Mn
	441[a]	388[a]	184[a]	302[a]	287[a]	509

	Fe	Co	Ni	Pd		
	520	557	505	452		
$M2O_3$	**Al**	**Ga**	**In**	**As**	**Sb**	**Bi**
	843	527	439	110	192	290
MO_2	**Zr**	**Hf**	**Si**	**Ge**	**Sn**	**Pb**
	805	872	589[a]	472[a]	536[a]	282[a]

Table 3.3. Sublimation (Upperlines) and Melting Heats (Lowerlines) of Elements (kJ/mol)

Li	**Be**	**B**	**C**	N	**O**	**F**	**Ne**	**H**	**He**
159.3	324	565	716.7	472.7	249.2	79.4	2.14	1.028	0.09
3.0	7.9	50.2	117	0.71	0.44	0.51	0.33	0.116	0.018
Na	Mg	Al	Si	P	S	Cl	Ar	–	–
107.5	147.1	330.9	450.0	316.5	277.2	121.3	7.70	–	–
2.6	8.5	10.7	50.2	0.66	1.72	6.40	1.18	–	–
K	Ca	Sc	Ti	V	Cr	Mn	Fe	Co	Ni
89.0	177.8	377.8	473.4	515.5	397.5	283.3	415.5	426.7	430.1
2.3	8.55	14.1	14.15	21.5	21.0	12.9	13.8	16.2	17.5
Cu	Zn	Ga	Ge	As	Se	Br	Kr	–	–
337.4	130.4	272.0	372	302.5	227.2	111.9	10.68	–	–
13.3	7.1	5.6	36.9	24.4	6.7	10.6	1.64	–	–
Rb	Sr	Y	Zr	Nb	Mo	Tc	Ru	Rh	Pd
80.9	164.0	424.7	610.0	733.0	659.0	678	650.6	556	376.6
2.2	7.4	11.4	21.0	30	37.5	33.3	38.6	26.6	16.7
Ag	Cd	In	Sn	Sb	Te	I	Xe	–	–
284.9	111.8	243	301.2	264.4	196.6	106.8	14.93	–	–
11.3	6.2	3.3	7.15	19.8	17.4	15.5	2.27	–	–
Cs	Ba	La	Hf	Ta	W	Re	Os	Ir	Pt
76.5	179.1	431.0	618.4	782.0	851.0	774	787	669	565.7
2.1	7.1	9.7	27.2	36.6	52.3	34.1	57.8	41.1	22.2
Au	Hg	Tl	Pb	Bi	Po	At	Th	Pa	U
368.2	61.4	182.2	195.2	209.6	147.0	97.2	602	563	533
12.55	2.3	4.15	4.8	11.1	12.5	–	13.8	12.3	9.14
–	Ra	Ac	–	–	–	–	–	–	–
–	159	406	–	–	–	–	–	–	–
–	7.7	12.0	–	–	–	–	–	–	–

3.7. EVAPORATION HEATS OF COMPOUNDS

The difference between the melting as well as sublimation heat, as previously stated, can be used to calculate evaporation heat. Table 3.5 includes information from the cited book as well as papers along with an analysis of measurements taken directly of H_v from 1910 to 2010 (Arshad & Beg, 1967). Since the evaporation heats of organic compounds are additive properties, they can be written as the sum of increments relating to various atoms, chemical bonds, and/or radicals (McEachern & Sandoval, 1973). The formula was developed by Korolev et al. (1997) as H_v(kJ/mol)=H_{vX}+H_{vCH}, where X and the C-H bond additions are represented by H_vX and H_{vCH}, respectively.

$$\Delta H_v\left(\text{kJ / mol}\right) = 4.69\left(n_C - n_Q\right) + 1.3n_Q + \text{a} + 3.0$$

where; 'a' represents an atomic group increment; 'n_C' is the overall amount of C atoms; and 'n_Q' is the total the amount of quaternary C atoms (Bolgar, Verkhoglyadova, & Samsonov, 1985). Table 3.3 displays the various functional groups' contributions to the evaporation enthalpy of molecules.

Enthalpy of vaporization appears to be a linear function of the number of carbon atoms, not only for alkanes or nitriles but also for alkyl-azides. The vaporization enthalpy (in kJ/mol) for mono-azides $CH_3(CH_2)_nN_3$ varies with the number of carbon atoms (N_C = 4–10) as well as follows the equation ΔH_v =16.6 +5.8 N_C (Jakob, Stucki, & Kuhn, 1995).

The equation ΔH_v =16.6 +5.8 N_C describes the relationship between the vaporization enthalpy and the amount of carbon atoms in diazides $N_3(CH_2)_nN_3$ with n = (3 to 6) ΔH_v=33.3+6.6N_C (Tables 3.4 and 3.5).

Table 3.4. Heats of Sublimation of Metal Chalcogenides (kJ/mol)

Comp.	ΔH_s	Comp.	ΔH_s	Comp.	ΔH_s
Cs_2Te	160	ZnSe	377	SnS	220
Cu_2Se	126	ZnTe	332	SnSe	214
CuS	316	CdSe	326	SnTe	222
CuSe	215	CdTe	293	PbS	233
Ag_2Se	110	GeS	167	PbSe	226
Ag_2Te	160	GeTe	197	PbTe	224

Table 3.5. Heats of Evaporation of Inorganic and Organo-Metallic Compounds (kJ/mol)

AH_n	ΔH_v	AF_n	ΔH_v	ACl_n	ΔH_v	ABr_n/AI_n	ΔH_v
HF	25.2	S_2F_2	14.9	S_2Cl_2	41.1	S_2Br_2	53.9
HCl	16.2	BeF_2	213.0	SCl_2	43.8	BBr_3	30.5
HBr	17.6	XeF_2	53.5	$BeCl_2$	105	$AlBr_3$	23.5
HI	19.8	BF_3	19.3	$ZnCl_2$	126	$GaBr_3$	38.9
H2O2	48.5	PF_3	16.5	$CdCl_2$	124.3	NBr_3	44.1
H_2O	40.6	AsF_3	20.8	$HgCl_2$	58.9	PBr_3	48.5
H_2S	19.5	ClF_3	27.5	$SnCl_2$	86.8	$AsBr_3$	41.8
H_2Se	19.7	BrF_3	47.6	$PbCl_2$	127	$SbBr_3$	59
H_2Te	19.2	CF_4	13.5	$CrCl_2$	197	$BiBr_3$	75.4
B_2H_6	14.3	SiF_4	15.4	$CoCl_2$	169.0	CBr_4	38.1
NH_3	22.7	SF_4	21.1	$NiCl_2$	147.2	$SiBr_4$	37.9
N2H4	41.8	SeF_4	46.4	BCl_3	23.8	$GeBr_4$	46.6
PH_3	14.6	TeF_4	34.3	$GaCl_3$	23.9	$SnBr_4$	43.5
P2H4	28.8	XeF_4	60.0	NCl_3	32.9	$TiBr_4$	44.4
AsH_3	16.7	VF_5	44.5	PCl_3	32.6	ZnI_2	117
SbH_3	21.3	NbF_5	52.3	$AsCl_3$	35.0	CdI_2	115
CH_4	8.2	TaF_5	56.9	$SbCl_3$	45.2	HgI_2	64.0
C2H2	16.3	PF_5	17.2	$BiCl_3$	72.6	PbI_2	104
C2H4	14.0	AsF_5	20.8	CCl_4	29.8	BI_3	40.5
C2H6	14.7	MoF_5	51.8	$SiCl_4$	29.9	AlI_3	32.2
SiH_4	12.1	BrF_5	30.6	$GeCl_4$	27.9	GaI_3	56.5
Si_2H_6	21.2	IF_5	41.3	$SnCl_4$	34.9	InI_3	95.4
GeH_4	14.1	ReF_5	58.1	$TiCl_4$	37.5	PI_3	43.9
Ge_2H_6	25.1	OsF_5	65.6	$ZrCl_4$	51.5	AsI_3	56.5
SnH_4	19.0	SF_6	17.1	VCl_4	41.4	SbI_3	68.6
C4H8	24.1	SeF_6	18.3	$TeCl_4$	77.0	BiI_3	77.7
C5H10	28.5	TeF_6	18.8	$NbCl_5$	52.7	SiI_4	50.2
C6H12	33.0	MoF_6	29.0	$TaCl_5$	54.8	GeI_4	64.2
C6H6	33.8	WF_6	25.8	$MoCl_5$	62.8	SnI_4	57.2
C7H14	38.5	OsF_6	28.1	WCl_6	52.7	TiI_4	58.5
C8H16	43.4	IrF_6	30.9	–	–	–	–
–	–	UF_6	29.5	–	–	–	–

$M(CH_3)_2$	ΔH_v	$M(CH_3)_3$	ΔH_v	$M(CH_3)_4$	ΔH_v	$M(C_5H_5)_2$	ΔH_v
Zn	30.0	B	22.0	Si	25.5	Cr	49.5
Cd	37.3	Al	40.0	Ge	27.8	Fe	47.3
Hg	35.6	Tl	39.2	Sn	32.1	Os	56.3
Te	35.1	As	28.3	Pb	36.9	Ru	53.6
–	–	Sb	30.0	–	–	–	–
–	–	Bi	35.5	–	–	–	–

3.8. ENTHALPIES OF PHASE TRANSFORMATIONS

Sublimation as well as melting are first order shifts in phase that alter both the structure as well as the aggregate state, in contrast to polymorphic modifications to crystals that only change structures inside the same aggregate state. The enthalpy for such transitions (H_{tr}) is consequently considerably lower than H_s and is comparable to H_m (Granasy et al., 1996).

Depending on where a specific species is situated on the phase diagram, these values vary considerably. While all else sums up, the heat of a phase transition depends on the depth of the essential rearrangement. Therefore, H_{tr} = 0.5, 2.0, 0.3, and 0.35 kJ/mol are required for the sphalerite to wurtzite change in CdS, CdTe, PbS, and PbTe, respectively; for the conversion of the same compounds to NaCl-type structures, H_{tr} = 20.6, 16.0, 2.3, and 3.0 kJ/mol are required, respectively (Ma & Atzmon, 1995). Reznitskii (2004) investigated the connection between the bonding of cations and the thermodynamic characteristics of oxides. He found a few associations in crystalline oxides as well as showed that it was possible to infer information about these substances' energetic properties from their structural makeup (Wang & Navrotsky, 2005). The enthalpy of crystals throughout a phase transformation can therefore range from about 1 to several 100 kJ/mol, according to the kind of structural changes, such as a change in the structure, the elimination of the short or distant order, or splitting into molecules or individual atoms in the gas phase (Chatterjee & Chakraborty, 2005) (Tables 3.6 and 3.7).

Table 3.6. Increasing of the Enthalpy of Evaporation of Organic Compounds (kJ/mol)

Class	**Group**	***a***	**Class**	**Group**	***A***	**Class**	**Group**	***a***
Acid	–COOH	38.8	Amide	–CONH	42.5	Ketone	>CO	10.5

Alcohol	–OH	29.4	Chlo- ride	–Cl	10.8	Nitrile	–CN	16.7
Alde- hyde	–CHO	12.9	Bro- mide	–Br	14.4	Nitro	$–NO_2$	22.8
Amide,pri Amide,sec	$–NH_2$ >NH	14.8 8.9	Iodide Ester	–I –COO	18.0 10.5	Nitro- gen Sulfide	=N–	12.2 13.4

Table 3.7. Enthalpies of Phase Transitions of Elements (kJ/mol)

M	ΔH_{tr}	ΔH_m	**M**	ΔH_{tr}	ΔH_m	**M**	ΔH_{tr}	ΔH_m
Ac[a]	–	10.8	Hf	6.7	27.2	Sc	4.0	14.1
Be[b]	6.1	7.9	La[a]	3.56	9.70	Sm[a]	3.27	12.4
C	2.1	117	Li	0.04	3.00	Sn	2.1	7.15
C60	6.5c,d	–	Na	0.03	2.60	Sr	0.75	7.40
C70	5.9[d]	–	Ni	0.58	17.5	Tb[a]	4.90	15.8
Ca	0.9	8.55	Mn	6.23	12.9	Th[a]	3.5	17.3
Ce[a]	3.06	8.64	Nd[a]	2.48	9.78	Ti	5.77	14.15
Co	0.45	16.2	Pa[a]	6.6	18.9	Tl	0.4	4.15
Fe	1.11	13.8	Pr[a]	3.12	10.0	Ua	7.47	15.9
Gd	3.94	13.74	S	0.40	1.72	Y	5.0	11.4
–	–	–	–	–	–	Zr	5.9	21.0

Activity 3.1

Enlist 10 polymorphic substances.

3.9. SUMMARY

A structure is thermodynamically inert if the free enthalpy of reaction ΔG for changing its structure into any other framework is positive. When the rate of change from one structure to another is negligible, a thermodynamically unstable framework may still exist. Isomers are molecules with a comparable composition but different structures. Polymorphism is the term used to describe the corresponding phenomenon in crystalline solids. The crystal-chemical method is unable to forecast the parameters of polymorphic modifications due to difficulties in accurately calculating bond ionicity and atom radii in polar bonds, but a comprehensive physical theory is still lacking. Phase changes in organic solids typically involve shifting the orientation of molecules within the crystal space and take place at milder thermodynamic condition changes. Amorphous solids are capable of changing their (short-range) atomic structure in a similar way to crystalline solids. Some substances exhibit crystalline polymorphism and polymorphism in combination. The difference among the melting as well as sublimation heat can be used to calculate the heat of vaporization. Sublimation as well as melting are firstorder shifts in phase that alter both the structure and the aggregate state, in contrast to polymorphic changes in crystals that only change structures inside the same aggregate state. The enthalpy for such transitions (H_{tr}) is consequently much lower than H_s and is comparable to H_m.

REVIEW QUESTIONS:

1. Write a detailed note on polymorphism.
2. What is phase transition?
3. Write a note on evaporation heats of compounds.
4. Describe in detail the melting heats of compounds.
5. Write a note on sublimation heats of elements.

REFERENCES

1. Alvarenga, M. E., Valdo, A. K. S., Ribeiro, L., Do Nascimento Neto, J. A., De Araujo, D. P., Da Silva, C. M., & Martins, F. T., (2019). Exploring the structural landscape of 2-(thiophen-2-yl)-1, 3-benzothiazole: High-Z′ packing polymorphism and cocrystallization with calix [4] tube. *Acta Crystallographica Section C: Structural Chemistry, 75*(6), 667–677.
2. Arancibia, L., Naspi, M., Pucci, G., Rodriguez, M., & Di Salvo, F., (2020). Revisiting the absolute chirality and polymorphism of (–)-Istanbulin A. *Acta Crystallographica Section C: Structural Chemistry, 76*(9), 914–920.
3. Arshad, M., & Beg, A., (1967). Correlation of the physical properties of organophosphorus compounds. I. Density, boiling point, and the heat of evaporation of phosphines. *Bulletin of the Chemical Society of Japan, 40*(1), 15–18.
4. Belenkov, E., & Greshnyakov, V., (2014). New polymorphic types of diamond. *Journal of Structural Chemistry, 55*(3), 409–417.
5. Bernstein, J., (1993). Crystal growth, polymorphism and structure-property relationships in organic crystals. *Journal of Physics D: Applied Physics, 26*(8B), B66.
6. Bernstein, J., (2011). Polymorphism– A perspective. *Crystal Growth & Design, 11*(3), 632–650.
7. Bolgar, A., Verkhoglyadova, T., & Samsonov, G., (1985). *Vapor Pressure and Evaporation Rate of Certain Heat-Resistant Compounds in a Vacuum at High Temperatures 1*(3), 2–10.
8. Boulineau, A., Sirisopanaporn, C., Dominko, R., Armstrong, A. R., Bruce, P. G., & Masquelier, C., (2010). Polymorphism and structural defects in $Li_2 FeSiO_4$. *Dalton Transactions, 39*(27), 6310–6316.
9. Brittain, H. G., (2012). Polymorphism and solvatomorphism 2010. *Journal of Pharmaceutical Sciences, 101*(2), 464–484.
10. Burdett, J. K., Hughbanks, T., Miller, G. J., Richardson Jr. J. W., & Smith, J. V., (1987). Structural-electronic relationships in inorganic solids: powder neutron diffraction studies of the rutile and anatase polymorphs of titanium dioxide at 15 and 295 K. *Journal of the American Chemical Society, 109*(12), 3639–3646.
11. Carson, A., Stranks, D., & Wilmshurst, B., (1958). The measurement of very low vapor pressures using radioactive isotopes: The latent heat

of sublimation of mercury diphenyl. *Proceedings of the Royal Society of London. Series A. Mathematical and Physical Sciences, 244*(1236), 72–84.

12. Castro, R. H., (2013). On the thermodynamic stability of nanocrystalline ceramics. *Materials Letters, 96*, 45–56.
13. Chatterjee, D., & Chakraborty, S., (2005). An enthalpy-based lattice Boltzmann model for diffusion dominated solid–liquid phase transformation. *Physics Letters A, 341*(1–4), 320–330.
14. Clegg, W., (2019). Crystallographic curiosities: Polymorphism and structures with Z′> 1. *Acta Crystallographica Section C: Structural Chemistry, 75*(7), 833–834.
15. Cruz-Cabeza, A. J., & Bernstein, J., (2014). Conformational polymorphism. *Chemical Reviews, 114*(4), 2170–2191.
16. Dworkin, A., & Bredig, M. A., (1968). Diffuse transition and melting in fluorite and antifluorite type of compounds. Heat content of potassium sulfide from 298 to 1260 K. *The Journal of Physical Chemistry, 72*(4), 1277–1281.
17. Ernst, E., (2001). *Family Polymorphism.* Paper presented at the ECOOP 2001—Object-Oriented Programming: 15th European Conference Budapest, Hungary, Proceedings 15.
18. Fazio, R., & Van Der Zant, H., (2001). Quantum phase transitions and vortex dynamics in superconducting networks. *Physics Reports, 355*(4), 235–334.
19. Ford, E. B., (1945). Polymorphism. *Biological Reviews, 20*(2), 73–88.
20. Fultz, B., (2020). *Phase Transitions in Materials*: Cambridge University Press.
21. Gerasimova, D., Saifina, A., Zakharychev, D., Vandyukova, I., Fayzullin, R., Kurbangalieva, A., & Lodochnikova, O., (2020). Packing polymorphism on the example of 5-hydroxy-1-(4-methylbenzyl)-3-chloro-4-[(4-chlorophenyl) sulfanyl]-1, 5-dihydro-2 H-pyrrol-2-one: A crystallographic, thermochemical, and spectroscopic study. *Journal of Structural Chemistry, 61*, 476–488.
22. Gholivand, K., Afshar, F., Shariatinia, Z., & Zare, K., (2010). Polymorphism for a novel phosphoramidate; NMR and X-ray crystallography. *Structural Chemistry, 21*, 629–636.
23. Ghouili, A., Brahmia, A., & Ben Hassen, R., (2015). Polymorphism in 3-acetyl-4-hydroxy-2H-chromen-2-one. *Acta Crystallographica*

Section C: Structural Chemistry, 71(10), 873–877.

24. Granasy, L., Kemeny, T., Oszlanyi, G., Bortel, G., Faigel, G., Tegze, M., & Janossy, A., (1996). Enthalpies of phase transformations in the alkali fulleride RbC60. *Solid State Communications, 97*(7), 573–578.
25. Greenwood, N., Perkins, P., & Twentyman, M., (1967). Heats of sublimation of the triphenyl derivatives of boron, aluminum, gallium, and indium. *Journal of the Chemical Society A: Inorganic, Physical, Theoretical*, 2109–2112.
26. Grimme, S., Huenerbein, R., & Ehrlich, S., (2011). On the importance of the dispersion energy for the thermodynamic stability of molecules. *ChemPhysChem, 12*(7), 1258–1261.
27. Groom, C. R., & Allen, F. H., (2014). The Cambridge structural database in retrospect and prospect. *Angewandte Chemie International Edition, 53*(3), 662–671.
28. Gschneidner Jr, K., (1969). On the interrelationships of the physical properties of lanthanide compounds: the melting point, heat of formation and lattice parameter (s). *Journal of the Less Common Metals, 17*(1), 1–12.
29. Hayashi, S., Ozawa, R., & Hamaguchi, H.-O., (2003). Raman spectra, crystal polymorphism, and structure of a prototype ionic-liquid [bmim] Cl. *Chemistry Letters, 32*(6), 498–499.
30. Hubbard, K., & Schlom, D., (1996). Thermodynamic stability of binary oxides in contact with silicon. *Journal of Materials Research, 11*(11), 2757–2776.
31. Hückel, W., (1950). *Structural Chemistry of Inorganic Compounds* (Vol. 2): Elsevier Publishing Company.
32. Jakob, A., Stucki, S., & Kuhn, P., (1995). Evaporation of heavy metals during the heat treatment of municipal solid waste incinerator fly ash. *Environmental Science and Technology, 29*(9), 2429–2436.
33. Khanmohammadi, M., Garmarudi, A., Moazzen, N., & Ghasemi, K., (2010). Qualitative discrimination between paracetamol tablets made by near infrared spectroscopy and chemometrics with regard to polymorphism. *Journal of Structural Chemistry, 51*(4), 663–669.
34. Korolev, S., Hsieh, J., Gauss, G. H., Lohman, T. M., & Waksman, G., (1997). Major domain swiveling revealed by the crystal structures of complexes of *E. coli* Rep helicase bound to single-stranded DNA and ADP. *Cell, 90*(4), 635–647.

35. Landau, L., (1936). The theory of phase transitions. *Nature, 138*(3498), 840–841.
36. Ludwig, J., & Nilges, T., (2018). Recent progress and developments in lithium cobalt phosphate chemistry-syntheses, polymorphism and properties. *Journal of Power Sources, 382*, 101–115.
37. Ma, E., & Atzmon, M., (1995). Phase transformations induced by mechanical alloying in binary systems. *Materials Chemistry and Physics, 39*(4), 249–267.
38. Malakhova, Y. A., Sukhikh, T., Rakhmanova, M., & Vinogradova, K., (2022). Effect of polymorphism on the luminescent properties on silver (I) nitrate complexes with 2-amino-5-phenylpyrazine. *Journal of Structural Chemistry, 63*(3), 485–500.
39. Mauzerall, D., (1960). The thermodynamic stability of porphyrinogens. *Journal of the American Chemical Society, 82*(10), 2601–2605.
40. McEachern, D., & Sandoval, O., (1973). A molecular flow evaporation apparatus for measuring vapor pressures and heats of sublimation of organic compounds. *Journal of Physics E: Scientific Instruments, 6*(2), 155.
41. Munshi, P., Venugopala, K., Jayashree, B., & Guru Row, T. N., (2004). Concomitant polymorphism in 3-Acetylcoumarin: Role of weak C–H⊙⊙⊙ O and C– H⊙⊙⊙ π interactions. *Crystal Growth & Design, 4*(6), 1105–1107.
42. Pappa, G. D., Voutsas, E. C., Magoulas, K., & Tassios, D. P., (2005). Estimation of the differential molar heat capacities of organic compounds at their melting point. *Industrial and Engineering Chemistry Research, 44*(10), 3799–3806.
43. Pettifor, D. G., (1989). New many-body potential for the bond order. Physical review letters, 63(22), 2480.
44. Politzer, P., Ma, Y., Lane, P., & Concha, M. C., (2005). Computational prediction of standard gas, liquid, and solid□phase heats of formation and heats of vaporization and sublimation. *International Journal of Quantum Chemistry, 105*(4), 341–347.
45. Politzer, P., Murray, J. S., Edward Grice, M., Desalvo, M., & Miller, E., (1997). Calculation of heats of sublimation and solid phase heats of formation. *Molecular Physics, 91*(5), 923–928.
46. Purohit, R., & Venugopalan, P., (2009). Polymorphism: an overview. *Resonance, 14*, 882–893.

47. Repická, Z., Moncol, J., Puchoňová, M., Jorík, V., Mikloš, D., Lis, T., & Valigura, D., (2010). [Cu (X-salicylato) 2 (N, N-diethylnicotinamide) 2($H_2O)_2$] complexes: conformational polymorphism and its consequence in supramolecular hydrogen-bonding networks formation. *Structural Chemistry, 21*, 1093–1102.
48. Schellman, J. A., (1987). The thermodynamic stability of proteins. *Annual Review of Biophysics and Biophysical Chemistry, 16*(1), 115–137.
49. Schmidt, J., Shi, J., Borlido, P., Chen, L., Botti, S., & Marques, M. A., (2017). Predicting the thermodynamic stability of solids combining density functional theory and machine learning. *Chemistry of Materials, 29*(12), 5090–5103.
50. Selbach, S. M., Einarsrud, M.-A., & Grande, T., (2009). On the thermodynamic stability of $BiFeO_3$. *Chemistry of Materials, 21*(1), 169–173.
51. Sproul, A. B., (1994). Dimensionless solution of the equation describing the effect of surface recombination on carrier decay in semiconductors. *Journal of Applied Physics, 76*(5), 2851–2854.
52. Tarnai, T., (2002). Polymorphism in multisymmetric close packings of equal spheres on a spherical surface. *Structural Chemistry, 13*, 289–295.
53. Tosi, H. L., (2002). *Multilevel Theory, Research, and Methods in Organizations: Foundations, Extensions, and New Directions 1*(3), 1–9.
54. Urusov, V. S., (1995). Semi-empirical groundwork of the bond-valence model. *Acta Crystallographica Section B: Structural Science, 51*(5), 641–649.
55. Vasiliev, A., & Laptash, N., (2012). Polymorphism of KNaNbOF 5 crystals. *Journal of Structural Chemistry, 53*, 902–906.
56. Vogel, A., & Nilges, T., (2021). Ion dynamics and polymorphism in Cu20Te11Cl3. *Inorganic Chemistry, 60*(20), 15233–15241.
57. Wang, M., & Navrotsky, A., (2005). $LiMO_2$ (M= Mn, Fe, and Co): Energetics, polymorphism and phase transformation. *Journal of Solid-State Chemistry, 178*(4), 1230–1240.
58. Zargar, V., Asghari, M., & Dashti, A., (2015). A review on chitin and chitosan polymers: Structure, chemistry, solubility, derivatives, and applications. *ChemBioEng Reviews, 2*(3), 204–226.

59. Zelaya, C. A., Stevens, E. D., & Dowd, M. K., (2010). 6, 6′-Dimethoxygossypol: molecular structure, crystal polymorphism, and solvate formation. *Structural Chemistry, 21*(1), 113–122.

CHAPTER 4

Properties of Organic Molecules

Contents

Unit Introduction 86
4.1. Organic Molecules 86
4.2. Functional Groups 89
4.3. Lewis Structure of Organic Molecules 95
4.4. Hybridization 96
4.5. Acid Base Reactions and K_a/Pk_a Values 102
4.6. Summary 106
Review Questions: 107
References 108

UNIT INTRODUCTION

The study of structure & behavior of organic molecules is an essential part or area of chemistry that has multiple applications in other fields such as medicine, materials science, and environmental science. The fields of health, environmental science, and materials science are included in this category. The most common type of atom found in organic compounds is carbon, however, these molecules may also contain atoms of hydrogen, oxygen, nitrogen, sulfur, or halogens. Carbon is the primary component of organic molecules. Compounds are what organic molecules are (Rosei et al., 2003).

Because these are some of the fundamental concepts in organic chemistry, they will be covered in this unit. Topics that will be covered include the structure and bonding of organic molecules, the characteristics of functional groups, and the processes that organic molecules go through (Alkorta et al., 1998).

Tip

Esters are organic substances that give many fruits their distinctive scent. For instance, the aroma of bananas is caused by isoamyl acetate, whereas the aroma of pineapples is caused by ethyl butyrate.

4.1. ORGANIC MOLECULES

Carbon may be used to create a wide variety of compounds, both synthetic and organic. These chemical compounds' structures are based on carbon rings or chains. A "family" of chemical molecules known as a homologous series all have the same functional group in common (Figure 4.1).

The fundamental units that make up an organic molecule's structure affect its properties. Atoms within a single molecule are joined by strong covalent bonds. Between molecules, there exist weaker intermolecular forces (Feist et al., 2018)

Particular in organic compounds, their chemical and physical properties vary based on their structure not merely on their molecular formula (Figure 4.2).

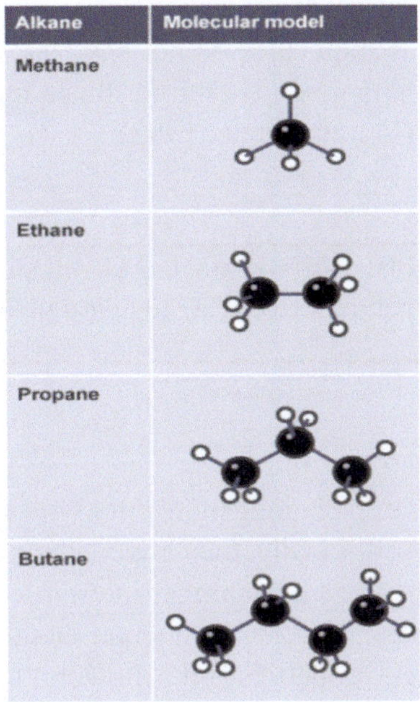

Figure 4.1. The structure of the alkanes is built on a carbon chain, and each carbon atom makes four covalent bonds.

Source: www.bbc.co.uk, Creative Commons License.

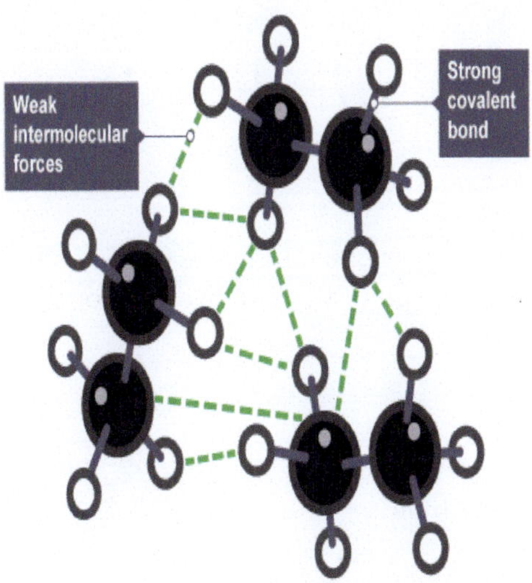

Figure 4.2. Strong covalent bonds and weaker intermolecular interactions holding together the atoms in molecules.

Source: www.bbc.co.uk, Creative Commons License.

One of the main difference between ionic compounds and organic molecules is that organic molecules do not have an overall electric charge, nor are their electrons unbound. Therefore, these materials don't conduct heat or electricity well (Eramian et al., 1993).

Did You Know?

Coffee, tea, and chocolate all contain the chemical substance known as caffeine. It works as a natural insecticide to guard plants from insect harm.

4.1.1. Polymers

Polymers are made of extremely long molecules with carbon as the backbone of the chain sequence. Additionally, they are connected by incredibly strong covalent bonds. Longer chains have higher intermolecular forces than more compact single molecules. These fundamental chain structures of organic molecules determine their properties and behavior. For instance, the melting point of polymers is higher than simple organic molecules due to the presence of longer stable non-polar chains (Beaujuge et al., 2010). The same complexity trend follows in terms of other properties such as boiling point, chemical reactivity and solubility to organic solvents (Figure 4.3).

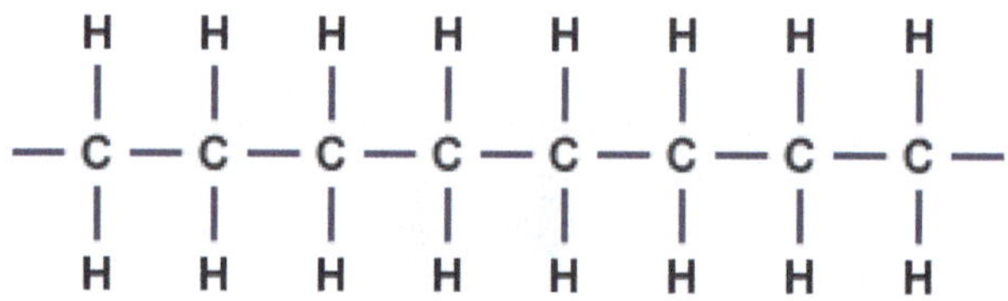

Figure 4.3. A model of a short segment of the basic polymer poly(ethene) molecule. Numerous chains made up of thousands of carbon atoms make up polyethene molecules.

Source: www.bbc.co.uk.

The intermolecular forces that exist among atoms of the organic molecules therefore affect their:

- Hardness;
- Melting point;
- Behavior on heating (whether polymers return easily to its original shape after heating);

- Flexibility;
- Solubility;
- Reactivity.

4.2. FUNCTIONAL GROUPS

Functional groups are usually substances that are found attached to the non-polar portion of organic molecules. When attached to a hydrocarbon chain, the functional groups allows the organic molecule to behave in unique and distinct way chemically. The functional groups also allow the easy classification or clustering of organic molecules. For instance, alcohols and carbonyl groups are the two most typical types of organic functional groups that you may encounter. Both alcohols and carbonyl groups group contain oxygen in the long hydrocarbon chain (Glusker, 1991).

The two groups of organic compounds mentioned above – alcohols, and carboxylic acids serve as specific examples of homologous sequences, often called families. Because they all contain the same specific set of atoms—their functional group—and molecules in a homologous series, they react very similarly to one another. Despite having varied carbon chain lengths or configurations, all members of a homologous series share the same functional group at some point in their structural makeup. This implies that they engage in similar broad responses (Ambrosetti et al., 2016).

Functional groups are helpful because they make it possible to infer properties of unknown molecules from their structure alone. If we can pinpoint its functional group, we may confidently predict how it will behave by comparing it to other molecules in the same homologous series or family (Perdue, 1985).

With the aid of a useful functional groups table, let's learn more about the numerous functional groups you'll face in organic chemistry (Table 4.1).

Table 4.1. The Functional Group Formula and an Example Molecule Along with the Functional Group Names Connected to Several Organic Families

Family Name	Functional Group Name	Functional Group Formula	Example
Alkane	C–C single bond	C–C	Ethane
Alkene	C=C double bond	C=C	Ethene
Alkyne	C≡C triple bond	C≡C	Ethyne
Alcohol	Hydroxyl	–OH	Methanol

Haloalkane or halogenoalkane (R=alkane; X=halogen)	Halogen atom	R-X	Chloromethane
Aldehyde	Carbonyl (aldehyde)	–CHO	Propanal
Ketone	Carbonyl (ketone)	–CO–	Propanone
Carboxylic acid	Carboxyl	–COOH	Ethanoic acid
Acyl chloride	Acyl chloride	–COCl	Ethanoyl chloride
Ester	Ester link	–COO–	Methyl ethanoate
Amine	Amine	$–NH_2$	Methylamine
Amide	Amide	$–CONH_2$	Ethanamide
Arene	Arene	C_6H_6	Benzene
Halogenoarene	Phenyl Halogen atom	C_6H_5X	Chlorobenzene
Phenol	Phenyl Hydroxyl	C_6H_5OH	Phenol (hydroxybenzene)

Source: bbc.co.uk.

4.2.1. Examples of Functional Groups

You may find the names of molecules that serve as excellent examples of functional groups in Table 4.1.

4.2.1.1. Alkanes, Alkenes, and Alkynes

You can see instances of the functional groups present in alkanes, alkenes, and alkynes in the first image. All three of these diverse groups of organic compounds are hydrocarbons, which means that what you will see are only atoms or carbon connected to other carbon atoms and hydrogen (Hambley, 1998). The carbon-to-carbon chain or bonds can be in terms of single bonds, double or triple bonds. When carbon to carbon bonds is single, they are called Alkanes, Alkenes and alkynes are the general term used for C=C double and C≡C triple bonds (Hyman et al., 1998).

Non-polar functional groups include the C-C single bond, C=C double bond, and C-C triple bond. This is due to the fact that the group's two constituent elements—carbon and hydrogen—have relatively comparable electronegativity values. For further information, visit Polarity and Electronegativity (Figure 4.4).

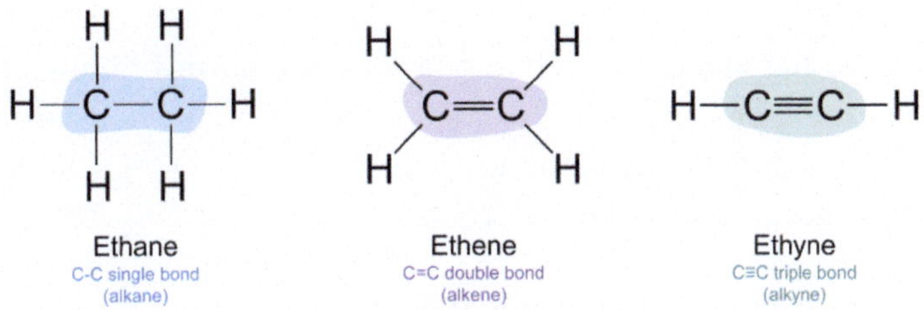

Figure 4.4. Alkane, alkene, and alkyne examples with emphasis on their functional groups. They are all hydrocarbon compounds.

Source: StudySmarter.

4.2.1.2. Alcohols and Halogenoalkanes

The hydroxyl functional group can be found in alcohols. A hydrogen atom is joined to an oxygen atom, which is subsequently joined to the remaining parts of the molecule. We use –OH to denote this category (Figure 4.5).

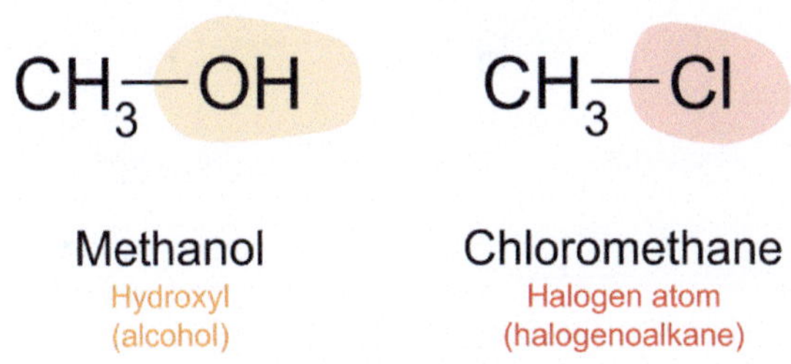

Figure 4.5. The hydroxyl group and a halogen atom are examples of functional groups that can be found in alcohols and halogenoalkanes, respectively.

Source: StudySmarter Original.

Even more straightforward, the halogen atom serves as the functional group in the haloalkane family. Typically, a halogen atom is represented by the symbol X, giving halooalkanes the functional group –X. Halogens and the hydroxyl group are both polar functional groups. Their polarity makes a wide variety of reactions possible, including nucleophilic substitution reactions. You'll learn about the hydroxyl group's ability to establish hydrogen bonds and how readily it can change into other functional groups in Reaction of Alcohols (Hermann et al., 2017).

4.2.1.3. Aldehydes and Ketones

All members of the aldehyde and ketone families have different forms of the carbonyl group, or C=O. An oxygen atom is connected to a carbon atom by a double bond to form the carbonyl group itself. The C=O can be found at the end of a carbon chain in the aldehyde functional group. This indicates that one R group and one hydrogen atom are connected to the carbon atom. The C=O is located in the center of a carbon chain in the ketone functional group. This indicates that two R groups are linked to the carbon atom. The only distinction between the ketone propanone (above, right) and the aldehyde propanal (above, left) is where the C=O carbonyl group is located. Aldehydes in addition are found at the extremes or end portion of the hydrocarbon chain, contrary to the ketone group that is found in the middle of a hydrocarbon chain (Hu et al., 2009). Aldehydes and ketones, however, have different characteristic reactions as a result (Figure 4.6).

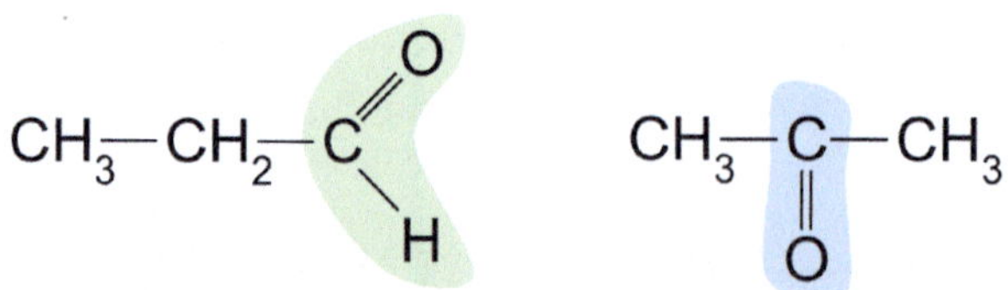

Propanal
Carbonyl
(aldehyde)

Propanone
Carbonyl
(ketone)

Figure 4.6. The carbonyl group is an illustration of a functional group present in aldehydes and ketones.

Source: StudySmarter Originals.

4.2.1.4. Carboxylic Acids, Acyl Chlorides, and Esters

The carboxyl functional group, or –COOH, is present in carboxylic acids. You can infer some details about this group's structure from its name. The two functional groups that make up this chemical are the carbonyl group (C=O, found in aldehydes and ketones) & hydroxyl group (–OH, found in alcohols). Because of the interaction of the two groups, carboxylic acids react by giving a proton in solution, just like other acids do (Figure 4.7).

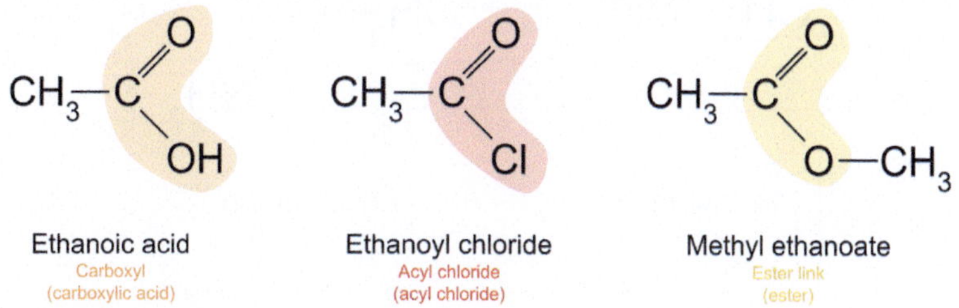

Figure 4.7. The carboxyl group, acyl chloride group, and ester link are examples of functional groups that can be found in carboxylic acids, acyl chlorides, and esters, respectively.

Source: StudySmarter Originals.

Esters and acyl chlorides are two different categories of acid derivatives. Acid derivatives are close relatives of carboxylic acids, in which an additional electronegative group has been added in place of the hydroxyl group (–OH). A chlorine atom serves as the replacement group in acyl chlorides, giving them the functional group –COCl (sometimes referred to as the acyl chloride group). The replacement group in esters is an oxygen atom bound in between two R groups. Esters are given the functional group –COO-, often known as the ester link, as a result ().

4.2.1.5. Amines and Amides

Like Oxygen and Hydrogen, Nitrogen in another element that has closer affinity to organic molecules as a member of a functional groupd. Our first illustration of an organic compound containing nitrogen is an amino (–NH_2) group, specifically known as amines. A nitrogen atom with a lone pair of electrons attached to at least one organic R group gives rise to the amine group. Amines react as bases because of the lone pair of electrons in nitrogen. Hydrogen bonds can also occur between amine groups that contain one or more hydrogen atoms. Amines can be primary, secondary, tertiary or quarternary, depending on the number of amino groups attach to the central carbon atom. Another group of organic molecules that contains nitrogen as a functional group are amides. The amide (–$CONH_2$) functional group, which itself is composed of an amine group (–NH_2) directly attached to a carbonyl group (C=O), defines them. They also create hydrogen bonds, just like amines (Figure 4.8).

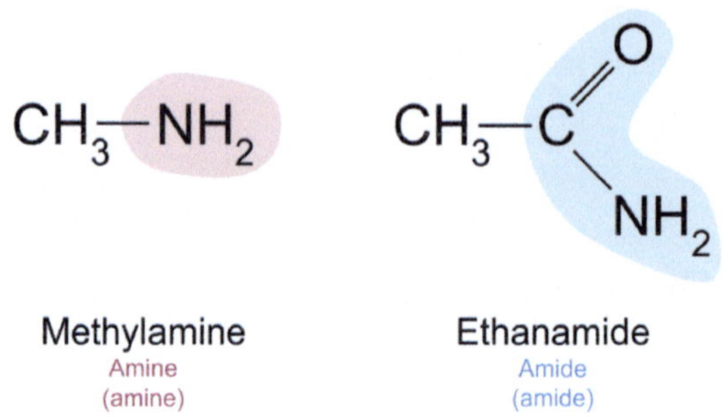

Figure 4.8. The amine group and the amide group are examples of functional groups in amines and amides.

Source: Studysmarter Originals.

4.2.1.6. Arenes, Halogenoarenes, and Phenol

The three organic families that follow are connected by an aromatic benzene ring. The arene functional group, sometimes referred to as benzene rings, has the chemical formula C_6H_6. To represent the arene group, however, using a hexagon with an inner circle, is the most commonly used structure representation for benzene (Figure 4.9).

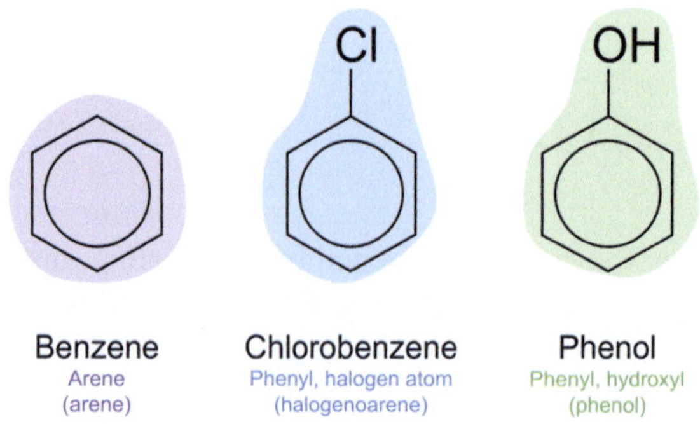

Figure 4.9. *Arenes, haloarenes, and phenols are a few examples of functional groups. An aromatic arene ring can be found in all three families. In contrast to phenols, which have a phenyl group attached to the hydroxyl group, haloarenes are composed of a phenol group connected to a halogen atom.*

Source: StudySmarter Originals.

Alkane, alkene, and alkynes of longer chain can form rings or cyclic groups that are unstable circles of delocalized pi electrons. Benzene rings contain a stable circle of delocalized pi electrons. Since this ring is robust and difficult to break, substitution reactions rather than addition reactions are more common for arenes and their derivatives. More information on this is provided in Benzene Electrophilic Substitution.

Haloarenes and phenols are examples of arene derivatives. A different group is substituted for one or more of the hydrogen atoms of the arene ring in arene derivatives. An arene ring without a hydrogen atom is referred to as a phenyl group. For instance, haloarenes combine the phenyl group and a halogen atom in its functional group by exchanging hydrogen atoms in the arene ring for halogen atoms. Phenols combine the phenyl group with the hydroxyl group (–OH) in their functional group by exchanging hydrogen atoms for the hydroxyl group (–OH) present in alcohols (Batsanov, 1999).

4.3. LEWIS STRUCTURE OF ORGANIC MOLECULES

The structure of organic molecules serves as the foundation for Lewis structures. They show how electrons are arranged in molecules, the bonding nature of each atom or molecule towards another in the electron level, specifically how they are shared between atoms to form particular bonds (Beaujuge et al., 2010).

4.3.1. Drawing Lewis Structure

So how does one go about creating an organic molecule's Lewis structure? Electrons, specifically valence electrons, are represented as dots or crosses around the atomic or molecular symbol when illustrating Lewis structures. Paired electrons are represented as lines to represent bonding, while unpaired and lone pairs of electrons are left as dots around the atom or molecule symbol (Bandyopadhyay & Sadhukhan, 2022).

You must first count the number of valence electrons that are present in your center atom. Draw the valence electrons as dots surrounding the atom's symbol. There is no prescribed placement of the electrons around the center atom which for ease of convention, the top-bottom-left-right process of distributing the dots around the atom becomes convenient when analyzing later electron sharing and bonding with another atom or molecule.

The second step is to identify the other atoms that are a part of the organic molecule you are attempting to portray and their valence electrons. More crucially, when all valence electrons are shown, you may make an educated prediction as to how electrons are exchanged between two atoms.

Show that two atoms are connected by their shared electrons (Dvorak et al., 1992). Finally, draw a line through all places where there are two electrons are shared or bonded. A bond is symbolized by a line that connects two elements. Draw two or three lines to represent a double or triple bond if two atoms share two or more pairs of electrons, such as three pairs. The location of a single pair of electrons is indicated by two dots next to one another.

4.3.2. Uses of Lewis Structure

Why are Lewis structures helpful, then? They are highly useful in figuring out how organic molecules are shaped.

Most notably, Lewis structures can clearly display two things: the number of bound electron pairs and lone electron pairs. These can establish the number of electron domains around each atom in the organic structure, which in turn establishes the overall structure of the molecule. This can be accomplished by interpreting the structure using the skeleton structures or the VSEPR theory (Gómez-Santos, 2009).

The Lewis structure in combination with curved arrows can illustrate the bonding process between molecules through electron sharing among organic molecules or electron transfer among ionic or inorganic compounds. The usual type of chemical reaction undergone by organic molecules are complex and majority are stepwise (multiple stages). The nature of these stepwise chemical reactions can be well understood by starting with the Lewis structure and arrows that will show the direction of electron sharing or delocalization in the electron level. This is true because electrons and even protons are usually transferred or shared one at a time in succession for each atom or molecule.

4.4. HYBRIDIZATION

To obtain the lowest energy state, atoms adopt hybridizations or redistribution of the energy orbitals to come up with new type of orbitals having entirely different energy and shape. The main goal is to distance bonds and lone pairs from one another as much as feasible in the absence of additional complicated circumstances (see below, for example, conjugation and aromaticity).

In the process, atomic orbitals of comparable energies merge or mix together forming new orbitals. In the *s,p,d,f* orbital orientation of an atom, for instance, the most usual merging is between the *s* and the *p* orbital giving way to the formation on a hybrid "*sp*" orbital and becomes quite useful in explaining the chemical activity and bonding, as well as the structure of atoms and molecules.

4.4.1. Rules for Hybridization of Atoms

Determine for each atom the correct Lewis structure the:

(σ bonds) + (lone pairs) = a number between 2 and 4

True for C/N/O and elements in the second row; higher for elements in the third row. Notice that it reads "lone PAIRS" and "SIGMA bonds"

(σ bonds) + (lone pairs)

= 4 → sp^3 hybridization = 4x sp^3 hybrid orbitals, tetrahedron, 109.5° bond angles

= 3 → sp^2 hybridization = 3x sp^2 hybrid orbitals and 1 p orbital, both trigonal planar and with 120° bond angles (the p orbital is parallel to the bonding plane), are used.

= 2 → sp hybridization = 2x sp hybrid orbitals and 2x p orbitals, linear, 180° bond angles (p orbitals perpendicular to bonding region and each other)

For ammonia (NH_3), in contrast to the ideal tetrahedron's 109.5° bond angle, the single pair on the N pushes the other three bonds away and reduces their angles to 107.3° (Figure 4.10).

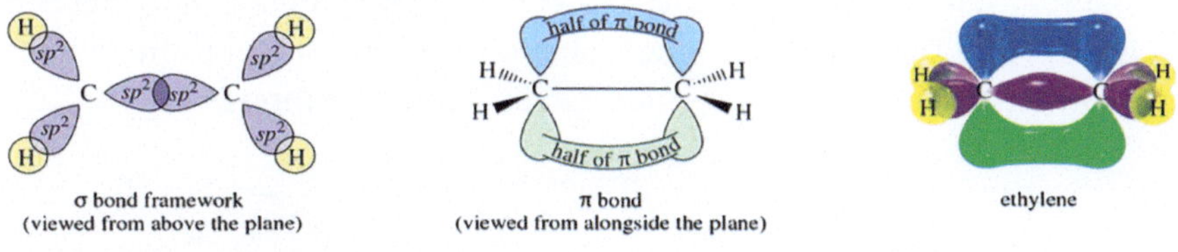

Figure 4.10. The remaining p orbital is used for the π bond.

Source: Pearson Education, Inc.

Acetylene C_2H_2

$$H - C \equiv C - H$$

The bonds that exist between C and H are sigma bonds and both carbon atoms are bonded to other 2 atoms (C and H). There is no lone pair of electrons as indicated by the Lewis structure. Thus, Carbon is sigma bonded to 2 other atoms, with no lone pairs which indicates an sp hybridization for the C atom in acetylene (Figure 4.11).

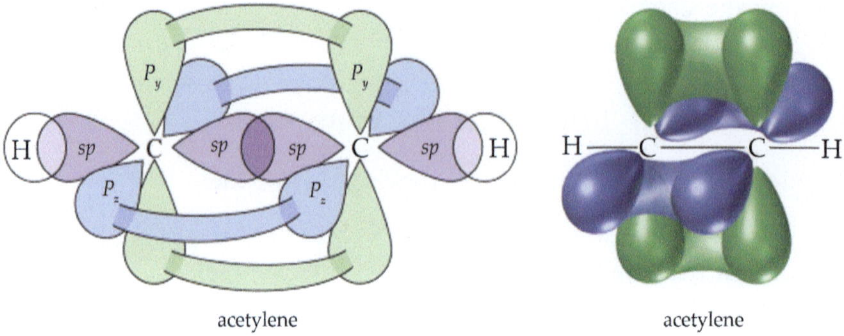

Figure 4.11. The remaining two p orbitals, form 2 perpendicular ð bonds.

Source: Pearson Education Inc.

4.4.2. Structure and Geometry

4.4.2.1. Rotation of Single Bonds

Structures with different *conformations* differ solely in rotation around a single bond. The sp^3 orbitals that overlap to create C-C bond are designed in a way that prevents interference from rotation about the bond axis. The bond here is fully rotatable and the bond strength remains constant throughout the rotation. The rotation in the sigma and single bonds are therefore characterized as "unrestricted" since the molecule can have infinite number of rotations which allows it to assume several conformations.

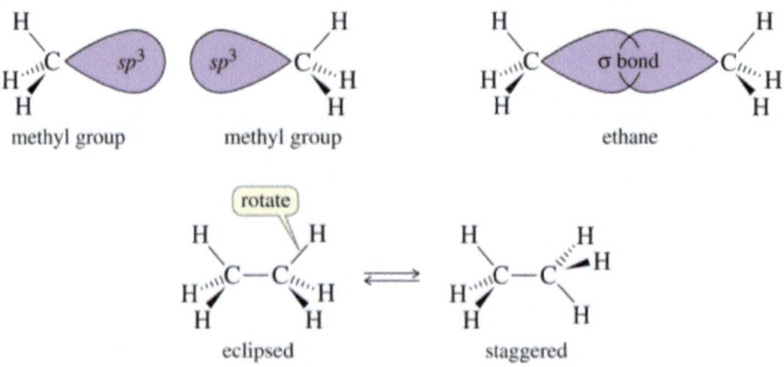

Figure 4.12. Tetrahedral and sp^3 hybridized carbons are both present in ethane. Ethane can exist in a variety of orientations by rotating one of the end methyl groups.

Source: Pearson Education Inc.

On the other hand, some conformations may have slightly higher energy content due to the different magnitude of attraction-repulsion forces between the atoms or groups near and around the carbon atom. The rotation in ethane molecule in Figure 4.12 illustrate this type of rotation (Jeffrey & Saenger, 2012).

4.4.2.2. Rigidity of Double Bonds

When the carbon atom in an organic molecule assumes multiple bonding in the case of double and triple bonds, rotation becomes limited and even restricted, which makes the molecule rigid. The rigidity is observed to intensify towards the triple bonds. Pi bonds exists among organic molecules with carbon atoms in double bonds as a result of overlapping p orbitals (Figure 4.13).

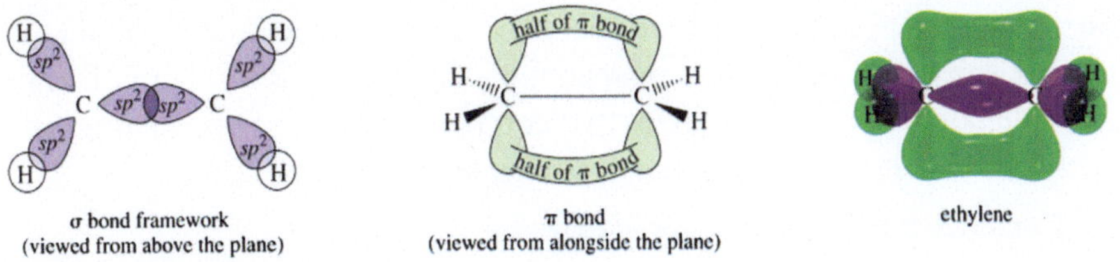

Figure 4.13. Sigma and Pi bond in ethylene.

Source: Pearson Education Inc.

The rotation of one of the CH_2 groups has no effect on the sigma bond because it is cylindrical. The difference is the pi bond. Rotation of one of the CH_2 groups would have negative impact on the overlap of the p orbitals. Rotation would cause a pi bond to rupture (Figure 4.14).

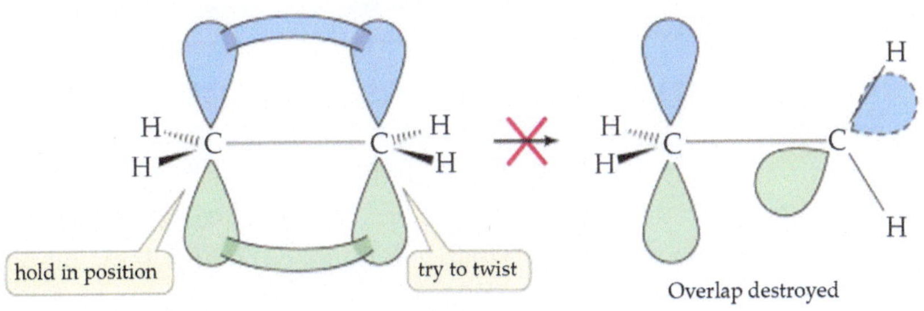

Figure 4.14. Pi bond break upon rotation of CH_2.

Source: Pearson Education Inc.

This follows that sigma and pi bonds in triple bonded carbons prevent bond rotation in such specific area. Thus, the triple bonds are more rigid than the double bond.

It is, therefore, safer to generalize that rotations in sigma or single bonded molecules are free and unrestricted while double and triple bonded molecules cannot rotate without breaking the pi-bonds that limits the rotation and makes the molecule rigid. The limiting nature of the pi-bonds on the sigma bonds in multiple bonded C atoms makes pi-bonds more exposed to potent reagents. It is the first one to be broken during chemical reactions, exposing the longer sigma bonds, which will later allow the molecules to be rotatable (Batsanov, & Batsanov, 2012).

4.4.2.3. Isomers

An isomer is a substance that shares a similar molecular formula with another substance but differs in spatial configuration or structure. Because a double bond cannot be rotated, two compounds can only be distinguished by the way their double bond substituents are arranged.

Consider but-2-ene (Figure 4.15):

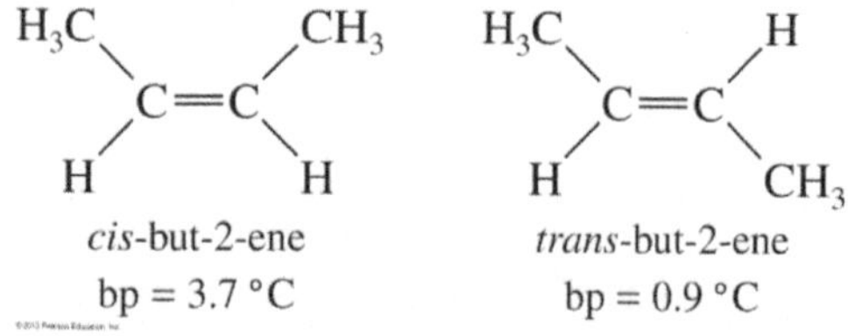

Figure 4.15. -cis-but-2-ene and trans-but-2-ene.

Source: Pearson Education Inc.

The methyl groups are on different sides of the double bond in the trans isomer of one molecule while it is located in the same side in the cis isomer. Rotation of the methyl group in the ends is prevented by the stiff double bond. They are two but-2-ene isomers. Stereoisomers are isomers that differ solely in how they are arranged spatially as opposed to how they are bound. Geometric isomers include cis and trans isomers. There must be two distinct groups on either end of the double bond for geometric isomerism to exist.

4.4.2.3.1. Structural Isomers

These are molecules that illustrates or exhibits different bonding pattern.

Several textbooks have been referring to the "connectivity of the C to C atoms" as the main driving force in the formation of structural isomers (Figure 4.16).

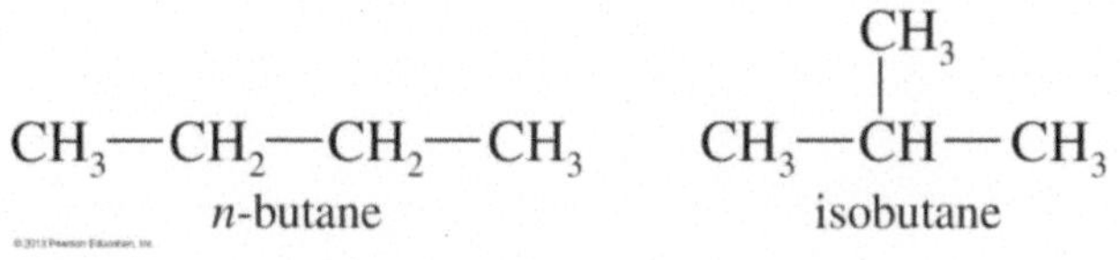

Figure 4.16. Structural isomers of n-butane and isobutane.

Source: Pearson Education Inc.

For instance, although having the identical structural formula of C_4H_{10}, the aforementioned compounds are different ones. The main difference therefore with n-butane and isobutane is that, after the 2nd Carbon atom from the left of n-butane, there attached $-CH_2$-CH_3, while there is $-CH_3$ attached to the 2nd C atom from the left of isobutane. Therefore, n-butane and isobutane are classified as constitutional or structural isomers.

Also, consider the 4 possible structural isomers ofC_5H_{10} below (Figure 4.17):

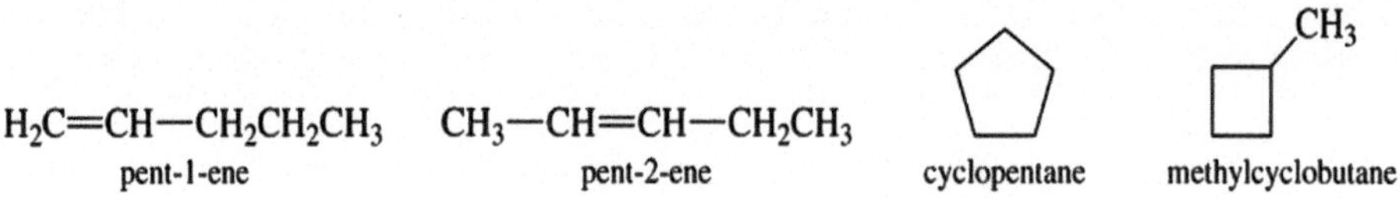

Figure 4.17. *Structural isomers of C_5H_{10}.*

Source: Pearson Education Inc.

4.4.2.3.2. Stereoisomers and Structural (Constitutional) Isomers

In general, isomers can either be stereoisomers and constitutional isomers. Stereoisomers are organic molecules having the same molecular composition but differ in 3D configuration. The cis- and trans-structure exhibited by several organic molecules as well as its handedness or chirality is generally under stereoisomerism, while, those that exhibits the same molecular make up but different C atom connectivities are classified as structural or constitutional isomers (Figure 4.18).

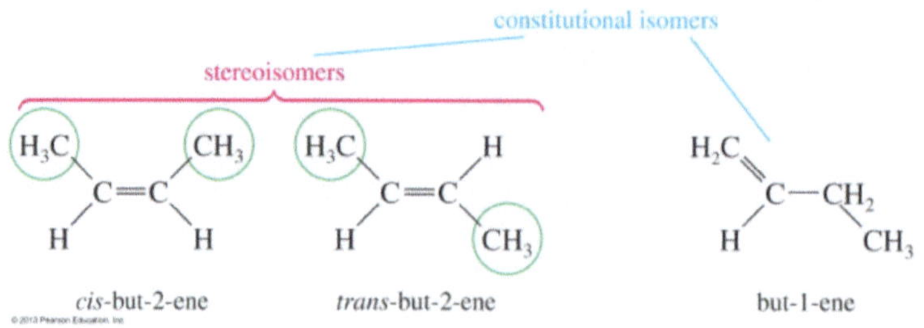

Figure 4.18. Stereoisomers and structural isomers.

Source: Pearson Education Inc.

4.5. ACID BASE REACTIONS AND K_A/PK_A VALUES

A very useful measure of acidity is the acidity constant K_a (and its negative logarithm, pK_a). The equilibrium constant for the dissociation of an acid into its conjugate base and a proton is represented by the acidity constant Ka.

Let us consider hydroiodic acid, or HI. HI will protonate water in a solution (let's use water) to produce H_3O^+ and I^-. The opposite reaction also takes place. [I^- and H_3O^+ react to produce HI as the byproduct]. This reaction is therefore in equilibrium.

By measuring the reaction's equilibrium constant, we can determine relative acidity of HI. Equilibrium constants can be determined directly through a series of experimental determination and also can be empirically deduced by related chemistry formula (Cole et al., 2004).

The reaction equilibrium constant for HI, therefore, turns out to be roughly 10,000,000,000. Accordingly, there are 10 billion molecules of I^- for every molecule of HI. In other words, HI, a prime example of a strong acid, rapidly gives its proton to give its corresponding base. Well, we can accomplish this for any species that contains a hydrogen, not only HI. By using the equilibrium constant or the "acid dissociation constant" is K_a, also considered as the *"ruler"* by which we can quantify any acid (Dryden et al., 2015). The K_a will allow us to measure the extent to which the acid dissociates in a solution and its resulting strength. The general convention on K_a is that the stronger the acid, the higher will be its K_a.

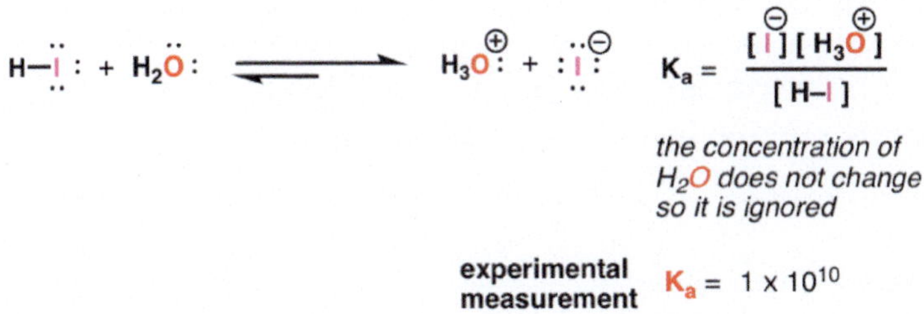

Acid dissociation examples constant, K_a for: extremely weak acid (CH_4), weak acid (CH_3OH), and a strong acid (HI)

Let's look at 3 examples. Note that these are somewhat oversimplified in that the solvent has been left out.

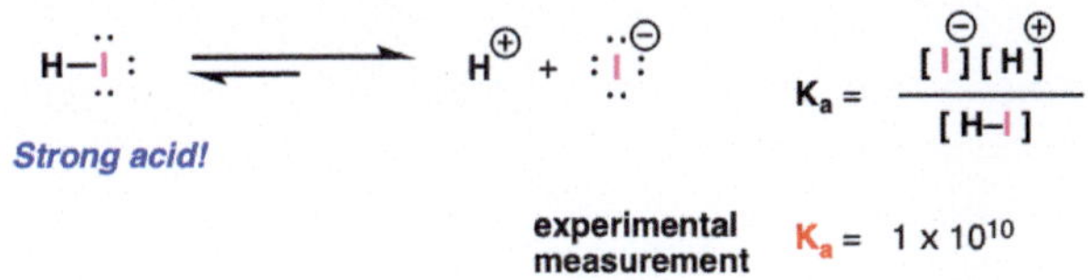

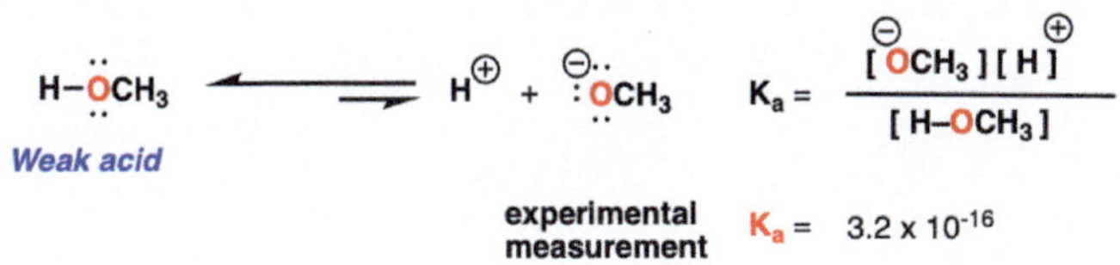

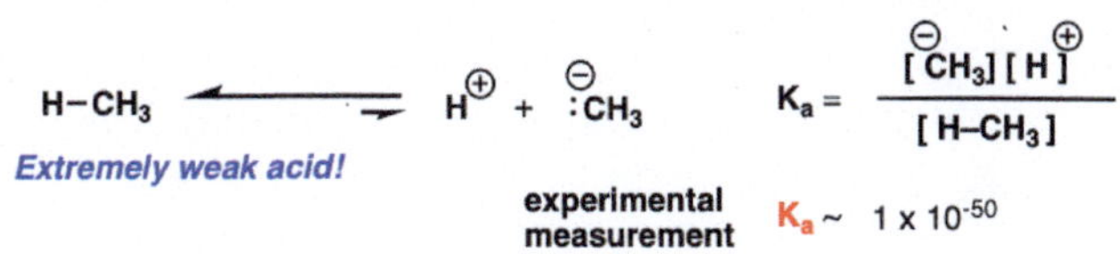

Now, measuring the acidity of numbers with many exponents after them is not the most practical method. So, we've switched to using a logarithmic scale instead. These are typical; for example, because the Richter scale is logarithmic, an earthquake with an 8.0 magnitude is 10 times more powerful than one with a 7.0 magnitude.

In comparison to K_a, the Negative Logarithm of K_a, or pK_a, is a far more practical indicator of acidity. More precisely, pK_a is the negative log base 10 of the K_a value which measures the strength of an acid – theoretically as to how tightly a proton is held by a Bronsted acid.

"pK_a" is the number we use to measure acidity. Since the vast majority of Ka values are below 1, and are more used to working with positive numbers, it is determined by taking the logarithm of the acidity constant K_a and randomly multiplying it by a negative number.

$$For\ H - I$$

$$Ka = 1x10^{10}$$

$$log\ Ka = 10$$

$$-log\ Ka = -10$$

$$This\ is\ the\ pKa\ of\ H - I\left(Strong\ Acid\right)$$

$$For\ H - CH_3$$

$$Ka = 1x10 -^{-50}$$

$$log\ Ka = -50$$

$$-log\ Ka = 50$$

$$This\ is\ the\ pKa\ of\ H - CH_3\left(weak\ acid\right)$$

Thus, the pKa for HI is –10 (signifying a very strong acid), for methanol it is 15 – weaker acid, and for methane is 50, signifying an incredibly weak acid. This further illustrates the relationship between K_a and pK_a which is inversely related. Simply put, the higher the Ka, the lower is the pK_a and the stronger is the acid.

The pKa scale encompasses over 60 orders of magnitude. That scale comprises 60 orders of magnitude.

The most potent among potent acids and the least potent among weak acids is readily illustrated in Table 4.2. Additionally, since the conjugate base is weaker when the acid is stronger, this also reveals information about basicity as well as conjugate acids and bases. In addition, Table 4.2 illustrates how a strong acid or bases tends to end up having weaker conjugates.

Table 4.2. The pKa Table Provides Details on the Breadth and Depth of a Variety of Chemical Behavior

Functional group	*Example*	*pKa*	*Conjugate Base*
Hydroiodic acid	H–I	–10	I^{-}
Hydrobromic acid	H–Br	–9	Br^{-}
Hydrochloric acid	H–Cl	–6	Cl^{-}
Sulfuric acid	HO–S(=O)₂–O–H	–3	HO–S(=O)₂–O^{-}
Hydronium ion	H_2O^{+}–H	–1.7	H_2O:
Sulfonic acids	H_3C–C₆H₄–S(=O)₂–O–H (tosic acid)	–1	H_3C–C₆H₄–S(=O)₂–O^{-}
Hydrofluoric acid	H–F	3.2	F^{-}
Carboxylic acids	H_3C–C(=O)–O–H	4	H_3C–C(=O)–O^{-}
Protonated amines	H–NH_3^{+} Cl^{-}	9-11	:NH_3
Thiols	CH_3S–H	13	CH_3S^{-}
Malonates	H_3CO–C(=O)–CH₂–C(=O)–OCH_3	13	H_3CO–C(=O)–CH^{-}–C(=O)–OCH_3
Water	HO–H	14	HO^{-}
Alcohol	H_3C–O–H	17	H_3C–O^{-}
Ketone/ aldehyde	H_3C–C(=O)–CH_3	20-24	H_3C–C(=O)–CH_2^{-}
Nitrile	H–C(H)(H)–C≡N:	25	H_2C^{-}–C≡N:
Ester	H_3CO–C(=O)–CH_3	25	H_3CO–C(=O)–CH_2^{-}
Alkyne	R–C≡C–H	25	R–C≡$C:^{-}$
Sulfoxide	H_3C–S(=O)–CH_3	31	H_3C–S(=O)–CH_2^{-}
Amine	H_2N–H	~35	:NH_2^{-}
Hydrogen	H–H	36	H^{-}
Alkene	H_2C=C(H)–CH_3	~43	H_2C=CH–C^{-}
Alkane	H_3C–CH_2–CH_2–CH_3	~50	H_3C–CH_2–CH_2–CH_2^{-}

Source: Master Organic Chemistry.

Did You Know?

Pure carbon that has been subjected to extreme pressure and heat for millions of years is what makes diamonds. Diamond is an exception to the rule of relatively stable organic compounds because of its distinctive crystal structure.

4.6. SUMMARY

Compounds with carbon atoms bound to other atoms, such as hydrogen, oxygen, nitrogen, sulfur, and halogens, are known as organic molecules. The configuration of these atoms dictates the behavior, reactivity, stability, and physical characteristics of the molecule.

The study of functional groups, which are certain groups of atoms found usually attached to non-polar organic molecule that govern a molecule's chemical behavior, is a significant area of organic chemistry. Alcohols, carbonyls, amines, and carboxylic acids are a few examples of functional groups.

Understanding the bonding arrangements and electron distribution in organic molecules can be done with the help of the Lewis structure. This entails creating diagrams that display the positioning of atoms and the shared as well as the lone pair of electrons among them.

Another crucial idea in organic chemistry is hybridization. It entrains the merging and overlapping of atomic orbitals to produce new hybrid orbitals that are used for chemical bonding. The shape and reactivity of the molecule are dependent on the type of hybridization.

A key concept in organic chemistry is the transport of protons (H^+) between molecules in acid-base reactions. Proton-donating organic compounds are known as organic acids, and proton-accepting organic bases are known as organic bases.

Overall, the study of the composition and characteristics of organic molecules is a challenging and intriguing area of chemistry with several real-world applications in fields including environmental research, materials science, and drug creation. Scientists can create novel compounds with certain features and uses by comprehending organic molecules' functional groups, Lewis's structure, hybridization, and acid-base interactions.

REVIEW QUESTIONS:

1. What is an organic molecule and what are some common functional groups found in organic compounds?
2. How are Lewis structures used to represent the bonding and structure of organic molecules?
3. What is hybridization and how does it influence the geometry and properties of organic molecules?
4. How do acid-base reactions occur in organic chemistry and what factors influence their strength and reactivity?
5. How do the functional groups present in an organic molecule influence its physical and chemical properties, such as boiling point, solubility, and reactivity?

REFERENCES

1. Alkorta, I., Rozas, I., & Elguero, J., (1998). Non-conventional hydrogen bonds. *Chemical Society Reviews, 27*(2), 163–170.
2. Ambrosetti, A., Ferri, N., DiStasio Jr, R. A., & Tkatchenko, A., (2016). Wavelike charge density fluctuations and van der Waals interactions at the nanoscale. *Science, 351*(6278), 1171–1176.
3. Andersson, Y., Hult, E., Rydberg, H., Apell, P., Lundqvist, B. I., & Langreth, D. C., (1998). Van der Waals interactions in density functional theory. *Electronic Density Functional Theory: Recent Progress and New Directions*, 243–260.
4. Bandyopadhyay, P., & Sadhukhan, M., (2022). Modeling coarse-grained van der Waals interactions using dipole-coupled anisotropic quantum Drude oscillators. *Journal of Computational Chemistry 3*(1), 7–17.
5. Batsanov, S. S., & Batsanov, A. S., (2012). *Introduction to Structural Chemistry*: Springer Science & Business Media *1*(2), 1–17.
6. Batsanov, S. S., (2000). Anisotropy of atomic Van der Waals Radii in the gas-phase and condensed molecules\sp {1}. *Structural Chemistry, 11*(2–3), 177–183.
7. Batsanov, S. S., (2001). Van der Waals radii of elements. *Inorganic Materials, 37*(9), 871–885.
8. Batsanov, S. S., Batsanov, A. S., Batsanov, S. S., & Batsanov, A. S., (2012). Intermolecular forces. *Introduction to Structural Chemistry*, 227–274.
9. Batsanov, S., (1995). Van der Waals radii of elements from the data of structural inorganic chemistry. *Russian Chemical Bulletin, 44*, 18–23.
10. Batsanov, S., (1999). Calculation of van der Waals radii of atoms from bond distances. *Journal of Molecular Structure: THEOCHEM, 468*(1–2), 151–159.
11. Batsanov, S., (2002). Correlation between the anisotropy of the electronic polarizability of molecules and the anisotropy of the Van Der Waals Atomic Radii. *Russian Journal of General Chemistry, 72*(8), 1153–1156.
12. Beaujuge, P. M., Amb, C. M., & Reynolds, J. R., (2010). Spectral engineering in π-conjugated polymers with intramolecular donor–acceptor interactions. *Accounts of Chemical Research, 43*(11), 1396–1407.

13. Beguin, L., Vernier, A., Chicireanu, R., Lahaye, T., & Browaeys, A., (2013). Direct measurement of the van der Waals interaction between two Rydberg atoms. *Physical Review Letters, 110*(26), 263201.
14. Bent, H. A., (1968). Structural chemistry of donor-acceptor interactions. *Chemical Reviews, 68*(5), 587–648.
15. Bondi, A. v., (1964). van der Waals volumes and radii. *The Journal of Physical Chemistry, 68*(3), 441–451.
16. Brown, I., (1976). On the geometry of O–H··· O hydrogen bonds. *Acta Crystallographica Section A: Crystal Physics, Diffraction, Theoretical and General Crystallography, 32*(1), 24–31.
17. Bullett, D., (1975). Chemical pseudopotential approach to covalent bonding. II. Bond lengths and bond energies in diamond, silicon and graphite. *Journal of Physics C: Solid State Physics, 8*(17), 2707.
18. Cole, M. W., Velegol, D., Kim, H.-Y., & Lucas, A. A., (2009). Nanoscale van der Waals interactions. *Molecular Simulation, 35*(10–11), 849–866.
19. Devi, T. G., Das, A., & Kumar, K., (2004). Raman anisotropic bandwidth study of C–O stretching vibration of methyl isobutyl ketone: role of van der Waals' volume of the interacting systems. *Spectrochimica Acta Part A: Molecular and Biomolecular Spectroscopy, 60*(1–2), 211–216.
20. Dryden, D. M., Hopkins, J. C., Denoyer, L. K., Poudel, L., Steinmetz, N. F., Ching, W.-Y., & French, R. H., (2015). van der Waals interactions on the mesoscale: Open-science implementation, anisotropy, retardation, and solvent effects. *Langmuir, 31*(37), 10145–10153.
21. Dunitz, J. D., & Gavezzotti, A., (2009). How molecules stick together in organic crystals: Weak intermolecular interactions. *Chemical Society Reviews, 38*(9), 2622–2633.
22. Dvorak, M., Ford, R., Suenram, R., Lovas, F. J., & Leopold, K. R., (1992). van der Waals vs. covalent bonding: Microwave characterization of a structurally intermediate case. *Journal of the American Chemical Society, 114*(1), 108–115.
23. Emmett, L., Prentice, G. M., & Pantoş, G. D., (2013). Donor–acceptor interactions in chemistry. *Annual Reports Section" B" (Organic Chemistry), 109*, 217–234.
24. Eramian, H., Tian, Y.-H., Fox, Z., Beneberu, H. Z., & Kertesz, M., (2013). On the anisotropy of van der Waals atomic radii of O, S, Se, F,

Cl, Br, and I. *The Journal of Physical Chemistry A, 117*(51), 14184–14190.

25. Feinberg, G., & Sucher, J., (1970). General theory of the van der Waals interaction: A model-independent approach. *Physical Review A, 2*(6), 2395.
26. Feist, J., Galego, J., & Garcia-Vidal, F. J. (2018). Polaritonic chemistry with organic molecules. ACS Photonics, 5(1), 205-216.
27. Fowkes, F. M., (1987). Role of acid-base interfacial bonding in adhesion. *Journal of Adhesion Science and Technology, 1*(1), 7–27.
28. Ganguly, P., & Desiraju, G. R., (2008). Van der Waals and polar intermolecular contact distances: Quantifying supramolecular synthons. *Chemistry – An Asian Journal, 3*(5), 868–880.
29. Glusker, J. P. (1991). Structural aspects of metal liganding to functional groups in proteins. Advances in protein chemistry, 42, 1-76.
30. Gobre, V. V., & Tkatchenko, A., (2013). Scaling laws for van der Waals interactions in nanostructured materials. *Nature Communications, 4*(1), 2341.
31. Gómez-Santos, G., (2009). Thermal van der Waals interaction between graphene layers. *Physical Review B, 80*(24), 245424.
32. Grabowski, S. J., (2006). *Hydrogen Bonding: New Insights* (Vol. 3): Springer.
33. Gregory, J., (1981). Approximate expressions for retarded van der Waals interaction. *Journal of Colloid and Interface Science, 83*(1), 138–145.
34. Gutmann, V., & Gutmann, V., (1978). *The Donor–Acceptor Approach to Molecular Interactions* (Vol. 228). Springer.
35. Hambley, T. W., (1998). van der Waals Radii of Pt(II) and Pd(II) in molecular mechanics models and an analysis of their relevance to the description of axial M⊙⊙⊙ H (– C), M⊙⊙⊙ H (– N), M⊙⊙⊙ S, and M⊙⊙⊙ M (M= Pd(II) or Pt(II)) interactions. *Inorganic Chemistry, 37*(15), 3767–3774.
36. Hancock, R. D., De Sousa, A. S., Walton, G. B., & Reibenspies, J. H., (2007). Metal-ion selectivity produced by C-Alkyl substituents on the bridges of chelating ligands: The importance of short H–H nonbonded Van Der Waals contacts in controlling metal-ion selectivity. A thermodynamic, molecular mechanics, and crystallographic study. *Inorganic Chemistry, 46*(11), 4749–4757.

37. Henrard, L., Hernandez, E., Bernier, P., & Rubio, A., (1999). van der Waals interaction in nanotube bundles: consequences on vibrational modes. *Physical Review B, 60*(12), R8521.
38. Hermann, J., DiStasio Jr, R. A., & Tkatchenko, A., (2017). First-principles models for van der Waals interactions in molecules and materials: Concepts, theory, and applications. *Chemical Reviews, 117*(6), 4714–4758.
39. Hobza, P., & Havlas, Z., (2000). Blue-shifting hydrogen bonds. *Chemical Reviews, 100*(11), 4253–4264.
40. Holstein, B. R., (2001). The van der Waals interaction. *American Journal of Physics, 69*(4), 441–449.
41. Hu, S.-Z., Xie, Z.-X., & Zhou, Z.-H., (2010). 70 years of crystallographic van der Waals Radii. *Acta Physico-Chimica Sinica, 26*(7), 1795–1800.
42. Hu, S.-Z., Zhou, Z.-H., & Robertson, B., (2009). Consistent approaches to van der Waals radii for the metallic elements. *Zeitschrift für Kristallographie, 224*(8), 375–383.
43. Hu, S.-Z., Zhou, Z.-H., Xie, Z.-X., & Robertson, B. E., (2014). A comparative study of crystallographic van der Waals radii. *Zeitschrift für Kristallographie-Crystalline Materials, 229*(7), 517–523.
44. Hyman, M. R., Murton, I. B., & Arp, D. J. (1988). Interaction of ammonia monooxygenase from Nitrosomonas europaea with alkanes, alkenes, and alkynes. Applied and environmental microbiology, 54(12), 3187-3190.
45. Jalenti, R., Lopez, A., & Azzi, A., (1980). Van Der Waals' volumes from densities and vaporization enthalpies. *Thermochimica Acta, 35*(3), 315–324.
46. Jeffrey, G. A., & Saenger, W., (2012). *Hydrogen Bonding in Biological Structures*. Springer Science & Business Media.
47. Kannemann, F. O., & Becke, A. D., (2010). van der Waals interactions in density-functional theory: Intermolecular complexes. *Journal of Chemical Theory and Computation, 6*(4), 1081–1088.
48. Kim, H., Cho, W. J., Madhav, M. V., & Kim, K. S., (2014). Anisotropic charge distribution and anisotropic van der Waals radius leading to intriguing anisotropic noncovalent interactions. *Scientific Reports, 4*(1), 1–8.
49. Kim, S. E., Mujid, F., Rai, A., Eriksson, F., Suh, J., Poddar, P., Zhong, Y., (2021). Extremely anisotropic van der Waals thermal conductors.

Nature, 597(7878), 660–665.

50. Kruse, H., Mrazikova, K., d'Ascenzo, L., Sponer, J., & Auffinger, P., (2020). Short but weak: The Z-DNA Lone-Pair··· π conundrum challenges standard carbon Van der Waals Radii. *Angewandte Chemie, 132*(38), 16696–16703.
51. Kuhn, L. P., (1952). The hydrogen bond. I. Intra-and intermolecular hydrogen bonds in alcohols. *Journal of the American Chemical Society, 74*(10), 2492–2499.
52. Kuz'mina, L. G., & Struchkov, Y. T., (1984). Structural chemistry of organomercury compounds. Role of secondary interactions. *Croatica Chemica Acta, 57*(4), 701–724.
53. Li, B., Yin, J., Liu, X., Wu, H., Li, J., Li, X., & Guo, W., (2019). Probing van der Waals interactions at two-dimensional heterointerfaces. *Nature Nanotechnology, 14*(6), 567–572.
54. Lin, I.-C., Seitsonen, A. P., Coutinho-Neto, M. D., Tavernelli, I., & Rothlisberger, U., (2009). Importance of van der Waals interactions in liquid water. *The Journal of Physical Chemistry B, 113*(4), 1127–1131.
55. Lin, Z., Huang, Y., & Duan, X., (2019). Van der Waals thin-film electronics. *Nature Electronics, 2*(9), 378–388.
56. Liu, L., Gong, P., Liu, K., Nie, A., Liu, Z., Yang, S., & Huang, L., (2022). Scalable van der Waals encapsulation by inorganic molecular crystals. *Advanced Materials, 34*(7), 2106041.
57. Morawietz, T., Singraber, A., Dellago, C., & Behler, J., (2016). How van der Waals interactions determine the unique properties of water. *Proceedings of the National Academy of Sciences, 113*(30), 8368–8373.
58. Nir, S., (1977). Van der Waals interactions between surfaces of biological interest. *Progress in Surface Science, 8*(1), 1–58.
59. Nyburg, S., & Faerman, C., (1985). A revision of van der Waals atomic radii for molecular crystals: N, O, F, S, Cl, Se, Br and I bonded to carbon. *Acta Crystallographica Section B: Structural Science, 41*(4), 274–279.
60. Pauling, L., & Beach, J., (1935). The van der Waals interaction of hydrogen atoms. *Physical Review, 47*(9), 686.
61. Perrin, C. L., & Nielson, J. B., (1997). "Strong" hydrogen bonds in chemistry and biology. *Annual Review of Physical Chemistry, 48*(1), 511–544.

62. Pimentel, G. C., & McClellan, A., (1971). Hydrogen bonding. *Annual Review of Physical Chemistry, 22*(1), 347–385.
63. Politzer, P., & Murray, J. S., (2021). The use and misuse of van der Waals radii. *Structural Chemistry, 32*, 623–629.
64. Perdue, E. M. (1985). Acidic functional groups of humic substances. Humic Substances in Soil, Sediment, and Water: Geochemistry, Isolation, and Characterization. John Wiley and Sons, New York NY. 1985. p 493-526, 8 fig, 1 tab.
65. Rosei, F., Schunack, M., Naitoh, Y., Jiang, P., Gourdon, A., Laegsgaard, E., ... & Besenbacher, F. (2003). Properties of large organic molecules on metal surfaces. Progress in Surface Science, 71(5-8), 95-146.
66. Roth, C. M., Neal, B. L., & Lenhoff, A. M., (1996). Van der Waals interactions involving proteins. *Biophysical Journal, 70*(2), 977–987.
67. Rowland, R. S., & Taylor, R., (1996). Intermolecular nonbonded contact distances in organic crystal structures: Comparison with distances expected from van der Waals radii. *The Journal of Physical Chemistry, 100*(18), 7384–7391.
68. Sandoghdar, V., Sukenik, C., Hinds, E., & Haroche, S., (1992). Direct measurement of the van der Waals interaction between an atom and its images in a micron-sized cavity. *Physical Review Letters, 68*(23), 3432.
69. Scheiner, S., (1997). *Hydrogen Bonding: A Theoretical Perspective*: Oxford University Press on Demand.
70. Steiner, T., & Desiraju, G. R., (1998). Distinction between the weak hydrogen bond and the van der Waals interaction. *Chemical Communications, 8*, 891–892.
71. Stöhr, M., Van Voorhis, T., & Tkatchenko, A., (2019). Theory and practice of modeling van der Waals interactions in electronic-structure calculations. *Chemical Society Reviews, 48*(15), 4118–4154.
72. Stone, A., (2013). *The Theory of Intermolecular Forces*: oUP oxford.
73. Strozzi, M., Smirnov, V. V., Pellicano, F., & Kovaleva, M., (2022). Nonlocal anisotropic elastic shell model for vibrations of double-walled carbon nanotubes under nonlinear van der Waals interaction forces. *International Journal of Non-Linear Mechanics, 146*, 104172.
74. Taylor, R., & Kennard, O., (1982). Crystallographic evidence for the existence of CH. N and CH. Cl hydrogen bonds. *Journal of the American Chemical Society, 104*(19), 5063–5070.

75. Vogt, J. r., & Alvarez, S., (2014). van der Waals radii of noble gases. *Inorganic Chemistry, 53*(17), 9260–9266.
76. Vold, M. J., (1961). The effect of adsorption on the van der Waals interaction of spherical colloidal particles. *Journal of Colloid Science, 16*(1), 1–12.
77. Weiss, M. S., Brandl, M., Sühnel, J., Pal, D., & Hilgenfeld, R., (2001). More hydrogen bonds for the (structural) biologist. *Trends in Biochemical Sciences, 26*(9), 521–523.
78. Wright, A. C., (2018). The structural chemistry of B_2O_3. *Physics and Chemistry of Glasses – European Journal of Glass Science and Technology Part B, 59*(2), 65–87.
79. Zefirov, Y. V., & Zorky, P. M., (1995). New applications of van der Waals radii in chemistry. *Russian Chemical Reviews, 64*(5), 415.

CHAPTER 5

Structure and Properties of Inorganic Chemistry

Contents

Unit Introduction 116
5.1. Introduction 116
5.2. Crystal Structures of Inorganic Compounds 119
5.3. Chemical Bonding 123
5.4. Ionic Bond 124
5.5. Covalent Bond 128
5.6. Physical Properties of Inorganic Compounds 133
5.7. Application of Inorganic Compounds 135
5.8. Summary 136
Review Questions: 137
References 138

UNIT INTRODUCTION

The carbon-hydrogen bonds characteristic of organic molecules is absent in inorganic substances, identifying them as distinct from organic molecules. Inorganic compounds can take the form of various substances, including metals, minerals, salts, and acids, amongst others.

The study of inorganic compounds encompasses a broad range of scientific disciplines, such as chemistry, physics, materials science, and engineering. Inorganic compounds are useful for various applications due to their unique properties, including electrical conductivity, magnetic susceptibility, optical activity, etc. Some examples of these applications include electronics, catalysis, and medicine (Bartram, 1990).

This chapter will discuss the structure and features of inorganic compounds. To begin, let's look at the essential components of inorganic substances, such as the atoms, ions, and molecules that make them up. Inorganic molecule creation and manipulation will also be covered in this course, along with topics such as chemical processes, crystal formation, and the deposition of thin films. In the final portion, we will discuss the significance of inorganic compounds in modern technology and industry and the challenges and opportunities facing inorganic chemistry.

Tip

Inorganic compounds, which are defined as those without carbon-hydrogen bonds, are often present in non-living materials including rocks, minerals, and metals.

5.1. INTRODUCTION

The study of the structure of chemical compoundss, discussed as stereochemistry or structural chemistry in the previous chapters, is a branch of chemistry that focuses primarily on molecular structures. In structural chemistry, spatial arrangement of atoms in a chemical is clarified and described, together with the driving forces and factors of this arrangement and the qualities that arise. Additionally, it includes a systematic grouping of known structure types & identification of connections between them. Structural chemistry is a crucial component of contemporary chemistry in both theory and practice. Knowing the structures of the chemicals involved is essential for adequate understanding of the processes occurring during chemical reactions & the creation of new molecular synthesis experiments. Understanding a substance's chemical and physical qualities requires knowledge of its structure (Aswathappa et al., 2020).

When comparing graphite and diamond, it is clear how much a material's structure affects its qualities. Although made entirely of carbon, they have different physical and chemical characteristics. The determination of structure is one of the most significant experimental jobs in structural chemistry. The most popular technique for doing this is X-ray diffraction from single crystals. However, other effective methods include neutron diffraction (ND) from single crystals & powders and X-ray diffraction from crystalline powders (Abdelrahman, 2023). An analytical component of structural chemistry involves structure determination, and the typical outcome is a static model. Experimentally, explaining spatial rearrangements of atoms during chemical reactions is significantly more difficult. This component of structural chemistry is addressed by the reaction processes relevant to the chemistry of molecules. Topotaxy studies chemical processes in solids, where structural relationships exist between orientation of reactants, intermediates & products. This book does not include dynamic features of this nature, experimental techniques for preparing solids, growing crystals, or figuring out structures. Crystals can be identified by their constituent parts' regular, periodic arrangement. Real crystals have many flaws, which become more prevalent as the temperature rises. Dislocations and other flaws can arise, such as missing or misaligned atoms. These flaws have a significant impact on a material's characteristics (Bekker et al., 2012).

5.1.1. Chemical Structures

We must adequately characterize the spatial distribution of the atoms to define complex chemical structures. The chemical nomenclature, which has been developed extensively, at least for the case of tiny molecules, can be of use in this regard. However, solid-state structures do not have a standardized nomenclature that allows us to express structural facts. The following method is utilized when handling the process of specifying the various types of structures: According to the assertion that "magnesium fluoride crystallizes in the rutile type," distribution of magnesium & fluorine atoms in magnesium fluoride (MgF_2) is identical to distribution of titanium & oxygen atoms in rutile (Abraham & Probert, 2006).

Graphical illustrations are helpful. One of these is a well-known valence-bond (VB) formula that makes it possible to describe a molecule's main structural elements briefly. Perspective, true-to-scale representations with the atoms represented as spheres or — if the continuously occurring thermal vibrations are to be portrayed — as ellipsoids are more accurate and illustrative. The ellipsoids or balls are presented on a smaller scale corresponding to an effective atomic size to provide a clearer picture. Covalent bonding is illustrated with sticks. The thermal ellipsoid size is set to match likelihood of discovering atom on average over time (often 50% probability of discovering atom's center within the ellipsoid; *see* Figure 5.1(b)). Drawings of many kinds can be used to highlight various construction features (Aswathappa et al., 2019).

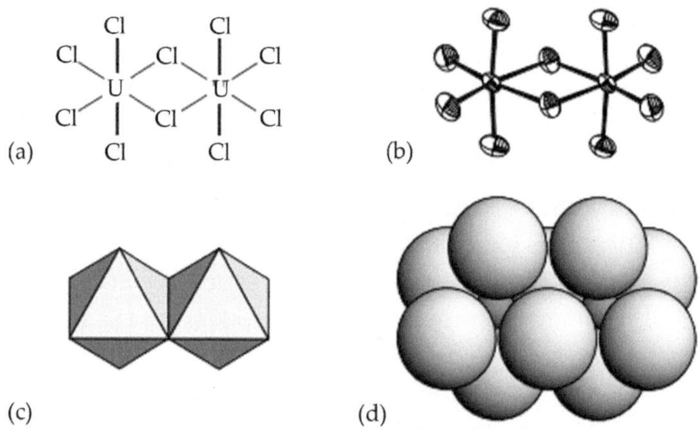

Figure 5.1. (a) A (UCl5-2) molecule is depicted graphically in identical scales. Valence-bond formula, in. (b) An elliptical perspective picture of thermal motion. (c) Polyhedra for coordination. And (d) emphasis on chlorine atom's need for space.

Source: Ulrich Muller, Inorganic Structural Chemistry.

Interatomic lengths and angles are given numerical values to create quantitative specifications. Interatomic distance is known as the distance between the two atoms' nuclei in their mean positions (thermal vibration's mean positions). Most popular technique for calculating interatomic distances experimentally is through x-ray diffraction from single crystals (Zvereva et al., 2006). Other techniques include microwave spectroscopy with gaseous samples, crystals ND, and electron diffraction for tiny molecules. Because X-rays are diffracted by electrons, electron shell centers rather than atomic nuclei positions are determined by X-ray diffraction. With the exception of hydrogen atoms that are covalently bound, the positions of the negative charge centers match the atomic nuclei very perfectly. Another process which is more preferrable than X-ray diffraction is ND because of its ability to pinpointing hydrogen atom locations. X-rays are diffracted much more strongly by numerous electrons of heavy atoms, making it difficult to pinpoint and establish the fix reliable position of hydrogen atoms. Neutrons, conversely, interact with atomic nuclei (Wherry, 1923). Therefore, this is not the case for them. (Neutron scattering is carried out with deuterated compounds because H atom nuclei inflict incoherent scattering on neutrons to a greater extent than D atom nuclei do).

Did You Know?

A lattice structure, in which ions are grouped in a repeating pattern, is frequently used to describe the chemical structure of inorganic compounds.

5.2. CRYSTAL STRUCTURES OF INORGANIC COMPOUNDS

5.2.1. Description

A crystal is a bigger network of connected atoms having a periodic order in three dimensions. The crystal structure provides the spatial arrangement of atoms in compound or molecule. A crystal lattice is formed by joining atoms of the same type that repeat in three directions on a 3-dimensional grid. Every lattice point is identical to every other point and is surrounded by the same elements (Wang et al., 2003). The crystal lattice can be imagined as being created by repeating little parallelepiped also called "unit cells" in three dimensions with no gaps (Figure 5.2; a parallelepiped is a body with six parallel faces).

Figure 5.2. Primitive lattice of cubic crystals. A unit cell.

Source: Ulrich Muller, Inorganic Structural Chemistry.

Three basis vectors with letters a, b, & c can be used to define the unit cell. Crystal lattice is a whole set of linear combinations $t = ua + vb + wc$, where u, v, w contains all positive & negative integers. "Crystal lattice" and "crystal structure" shouldn't be used interchangeably because the former refers to a physical structure, and the latter is an abstract geometric design (Tilley, 2006; Teitelbaum et al., 2018). A lattice parameter (or lattice constants; between a and b, etc.) is the length of basis vectors, a, b, & c, and the angles between them. There's no special method for selecting unit cells for specific crystal structures, as seen in Figure 5.3 for a 2-dimensional example. Some agreed-upon norms for the choosing of unit cell have been developed in crystallography to achieve consistency in crystal structures description (Figure 5.4):

- Basis vectors for unit cells must be perpendicular to symmetrical planes or parallel to symmetry axes in order to establish the crystal's symmetry.
- An inversion center is chosen as the geometrically distinct location for the origin of the unit cell (Swaminathan et al., 1984).
- It is best if the basis vectors are as brief as feasible. Angles between cells should be as near 90°, and cell volume should be as tiny as possible.
- Angles between the basis vectors are chosen to be greater than or equal to 90° (ideally greater than 90°) if the angles differ by degrees (Sørensen & Mølmer, 1999).

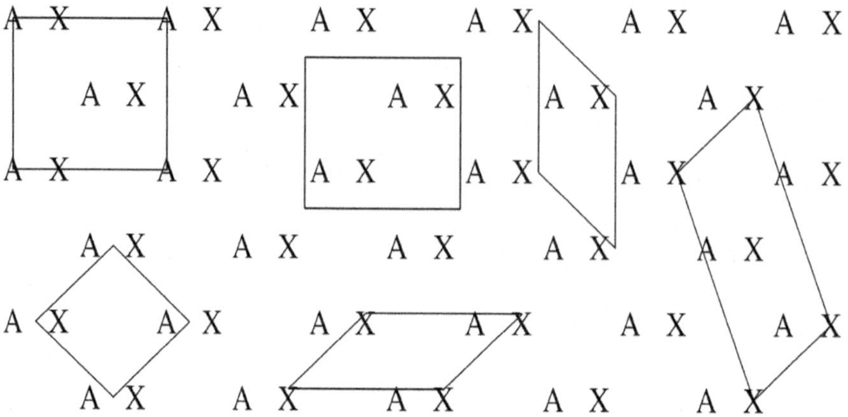

Figure 5.3. Atoms A and X are positioned in a regular, two-dimensional pattern. Any one of the depicted unit cells can be repeated to create the entire pattern.

Source: Ulrich Muller, Inorganic Structural Chemistry.

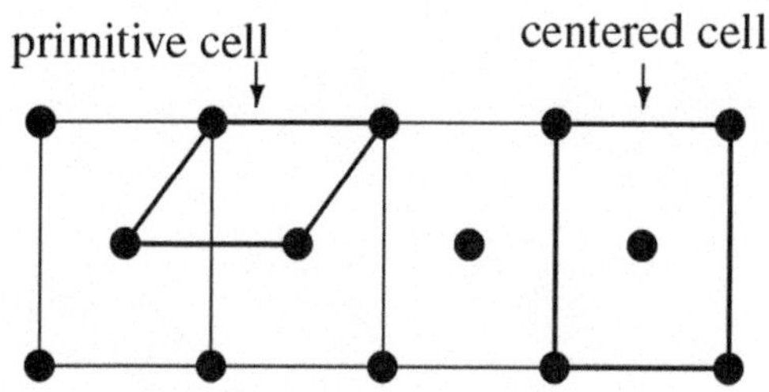

Figure 5.4. Primitive and centered cell.

Source: Ulrich Muller, Inorganic Structural Chemistry.

Primitive cells are the smallest possible unit of a cell. For symmetry reasons, centered cells that are double, triple, or fourfold primal, meaning their volume is greater by a corresponding factor, are usually chosen (Sivakumar et al., 2021). This conforms with rule 1 and goes against rule 3. Figure 5.5 depicts the centered cells to be considered.

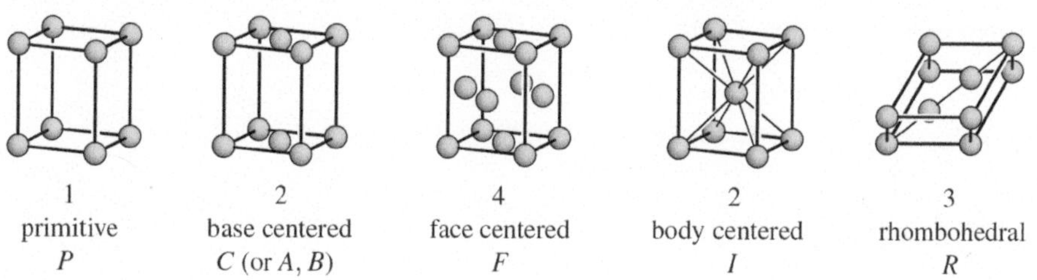

Figure 5.5. Unit cells that are centered and their symbols. Numbers indicate how numerously primitive each is.

Source: Ulrich Muller, Inorganic Structural Chemistry.

To establish uniform descriptions of crystal structures, additional guidelines have been created in addition to the norms specified for cell selection (Sivakumar et al., 2021). They must adhere to guarantee methodical & comparable documentation of data & to make database inclusion easier. Contraventions of standards, however, are rather common; they frequently result not only from carelessness or a failure to understand the norms but also from strong justifications, such as when it is necessary to draw attention to the connections between various structures (Showrilu et al., 2021). Only lattice parameters & positions of all atoms in the unit cell are required to characterize all the essential features of crystal structure. Only an integral number of atoms can fit inside of a unit cell. When specifying contents of a unit cell, quantity Z is commonly used to denote the number of "formula units" present. Atomic number counting is depicted in Figure 5.6.

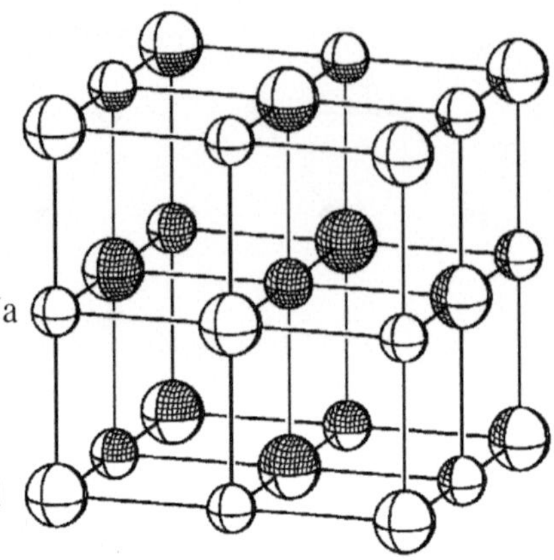

Figure 5.6. For face-centered unit cell of NaCl, here is how to count the contents: 12 Na ions in the centers of 12 edges, each of which belongs to 4 cells, makes 124 3; 1 Na ion in the center of the cube, which belongs only to this cell, makes 88 1. There are also 6 Cl ions in the centers of 6 faces, each belonging to two adjacent cells. Total: 4 units of the formula NaCl (Z = 4) or 4 Na and 4 Cl ions.

Source: Ulrich Muller, Inorganic Structural Chemistry (2006).

5.2.2. Lattice Energy

The energy generated during the assembly of a single mole of a crystalline substance from its infinitely spaced-apart components at a temperature of 0 K is known as lattice energy (Shilkina et al., 2020).

Components in this sense are understood to be:

- In the case of molecules;
- The ionic compound molecules;
- Atoms for pure elements, eliminating molecular species like H_2, N_2, S_8, etc.; ions for metals; the elemental form.

Only compounds that cannot be categorically classified into one of these substance classes when kind of components is known are appropriate candidates for defining a lattice energy. SiO_2, Si, and O atoms or Si_4 & O_2 ions should be employed. The lattice energy values provided in literature for polar compounds like SiO_2 typically refer to a composition of ions (Seitz, 1958; Puska, 1987). Values derived from this supposition should be interpreted cautiously because the omitted covalent bonds play a significant role. Conditions for an assembly from ions are not always evident, even in this case: should Na_2SO_4 be formed from Na & SO_4 ions or from Na, S_6, & O_2 ions?

5.3. CHEMICAL BONDING

5.3.1. Chemical Bonding and Structure

The distribution of their electrons considerably impacts which spatial arrangement of atoms lead to stable and metastable structures. The electrical arrangement of a single atom in a noble gas is thermodynamically stable under typical circumstances. It is still possible to release a little amount of *van der Waals energy* by simply compressing the atoms to create a liquid or solid. Due to the very low condensation and crystallization enthalpies and low melting and boiling points, the term TS already limits G size at relatively low temperatures (Roth, 2000). For all other elements, a single atom's electrical configuration does not, under typical circumstances, equate to a thermodynamically stable state. Single atoms don't exist in the vapor phase unless temperatures are high. Atoms must be connected to create stable structures at more typical temperatures. Like in a single atom, electrons in a group of atoms can only exist in specific, defined energy levels (Poulain et al., 2014).

The eigenvalues of wave functions can be used to represent these states numerically. Theoretically, the wave functions are the *Schrodinger* equation's solutions for an entire set of each constituent atom. We do have a solid understanding of wave functions and, consequently, of electrons in atomic systems, despite fact that finding the precise mathematical solution to this equation offers insurmountable challenges (Pekov et al., 2018). The information is based on mathematical approximations that have become increasingly dependable over time and experimental evidence. Ionic and covalent bonds, two extreme types of chemical bonding, is usually the starting point for a simplified model. However, the presence of the two bonding types must be considered to provide a range of middle states between these two extreme cases.

The weaker ion-dipole, dipole-dipole, and dispersion interactions must also be considered, to the extent that they are relevant forces to the vdW energy. The narrow range of action of the (localized) covalent bond, which typically only extends from one atom to the next, sets it apart from other types of bonds (Parker & Walker, 2002). However, it has a strong connection within this range. An atom develops a near order influenced by the close interatomic connections, mutual repulsion of valence electrons, and spatial needs of bound atoms. Similar to how near order of surrounding a brick propagates to a long-range order in a brick wall, when atoms are united to form larger structures, the near order can result in a long-range order (Panchadhyayee, 2018).

Covalent bonds in a non polymer molecule or ion connect only a few atoms. A molecule's internal covalent forces are much greater than any external forces. Because of this, when thinking about molecular structures, acting as if the molecule would arise by itself and have no surroundings is merely a minor mistake (Nieminen & Manninen, 1979). Bond lengths and angles in molecules often only experience minor changes as the molecules combine into a crystal, according to common experience, more thorough research by Kitaigorodsky (1978), and calculations with force fields. In some circumstances, only conformation angles are more strongly impacted. Therefore, molecular structures can be used to describe various features of molecular composition.This is not equally true for macromolecular compounds, where a molecule can have almost any number of atoms. In this instance, it is important to consider the interactions of nearby molecules. For a substance made of thread-like macromolecules, the physical characteristics depend on whether molecules are arranged in crystalline structure or tangled (Nelson & Toner, 1981). According to connectivity type of covalently connected atoms, crystalline macromolecular compounds can be divided into three categories: chain structures, layer structures, and framework structures.

Chains or layers could be polyanions or polycations held together by supplementary counterions or electrically neutral molecules that only interact with one another via van der Walls forces (Natta et al., 1965). Framework structures can also be charged, and counter-ions fill up any gaps in the network. The structure of a layer, chain, or framework is significantly affected by covalent bonds and tight order that results surrounding each atom. The crystal structure of small molecules ion-containing compounds are mostly determined by the ions ability to occupy available space efficiently while maintaining the idea of cations surrounding anions and anions surrounding cations. It is crucial to consider geometric elements like the relative ion sizes and molecular ion shapes (Murr et al., 1989).

5.4. IONIC BOND

An ionic bond can form when two or more atoms transfer electrons to form one atom to another. Ionic bonds are formed when a metal loses electrons & combines with a nonmetal gaining electrons. Ion bonds are created when opposing charges of ions are attracted to one another. These bonds are more powerful than hydrogen bonds but less powerful than covalent ones (Mestres et al., 1987; Li et al., 1993).

Covalent bonds between atoms include the sharing of electrons, whereas ionic bonds involve the attraction of opposing ions. Covalent bonding is governed by the valence shell electron pair repulsion theory (VSEPR rules), whereas maximal packing principles govern ionic materials. Therefore, the atomic configuration determines whether a molecule is ionic or covalent. It only occurs if the total energy shift of the reaction is is favorable since linked atoms possess lower energy than free ones. When the energy shift is higher, the link is stronger. Real atoms cannot form pure ionic bonds between atoms. Every bond has a small amount of covalence (Luu et al., 2020). The higher the electronegative difference, the higher the electrostatic attraction between atoms and the more ionic the binding. An idea of how an ionic connection would develop between two ions, such as $[Na]^+$ and $[Cl]^-$. The bond is exclusively (ideally) based on electrostatic interactions between positive and negative ions, and electron orbital typically does not overlap because each ion has reached its lowest energy state (i.e., a molecular orbital is not created). Majority of ionic compounds are soluble in water, while exceptions exist. The answer hinges on whether or not the water molecule attraction to the ions is stronger than the ion-ion attraction. The attraction of positive ions to the ion pairs on water molecules can lead to the formation of coordinated (dative covalent) bonds. Hydrogen bonds are formed between water molecules with the help of negative ions (Lennard-Jones & Devonshire, 1939).

5.4.1. Ionic Compounds

Ionic compounds, often known as salts, are made up of separate ions held together by electrostatic forces of attraction between cations and anions. The ionic compound characteristics provide insight into how ionic bonding is structured. Indicators of the strength of the ionic bonding include the crystalline structure, stiffness, and brittleness of ionic solids and their propensity for high melting and boiling temperatures. Ionic solids are poor electrical conductors because the strength of ionic connections prohibit ions from readily flowing through the solid state. Water is a good solvent for the vast majority of ionic substances. Once dissolved or molten, the ionic compound free-moving ions make them great heat and electricity conductors (Larson et al., 1964).

Ions connected to neutral atoms exhibit various physical and chemical characteristics. Sodium metal, a delicate silvery-white metal that burns violently in air & interacts vigorously with water, is created when sodium atoms combine. Chlorine atoms produce chlorine gas, often known as Cl_2, which is a highly toxic yellow-green gas that unsafe to both plants and animals as well as the majority of metals. Sodium chloride, usually referred to as common table salt, is a white, crystalline product resulting from the combination of elements sodium and chlorine. Sodium cations and chloride anions make up its composition (Lorenz & Whittaker, 1867). These ions combine to generate a substance with fundamentally distinct characteristics from sodium and chlorine. While chlorine is deadly and sodium metal is violently reactive with water, their combination produces sodium chloride – water soluble substance that, since its discovery is very vital for life (Figure 5.7).

(a) (b) (c)

Figure 5.7. (a) To prevent interaction with air or water, sodium, a soft metal, must be kept in mineral oil while not in use. (b) Chlorine gas has a vivid yellow-green color. (c) They combine to create white crystals of sodium chloride or table salt.

Source: modification of work by "Jurii"/Wikimedia Commons.

5.4.1.1. Ionic Compounds Formation

Binary ionic compounds only include two elements: metal (which forms cations) and nonmetal (which forms anions). NaCl is an illustration of a binary ionic compound. A periodic feature of the elements can be used to explain how such compounds are created. Numerous metallic elements easily lose electrons and have low ionization potential. In the periodic table, except hydrogen, these substances are located from the extreme left side to the far middle right portion of the periodic table. On the other hand, non-metals are found in the right side of the periodic table. Due to the relatively high electron affinities of non-metallic atoms, it can easily absorb electrons that metal atoms have lost, filling their valence shells (Kröger & Vink, 1956).

The total positive charge on the cations of an ionic compound must be equal to the total negative charge on the anions for the resulting compound to be electrically neutral. The simplest ratio of ions needed to produce same number of positive and negative charges is represented in the chemical formula for the ionic compound (Koshlyakova et al., 2021).

In the case of aluminum oxide, for instance, the formula Al_2O_3 shows that there are two aluminum cations, Al^{3+}, for every three oxide anions, O^{2-} [the total charge of the compound must be zero, therefore, $(2*(+3)) + (3*(-2)) = 0$].

But it's vital to remember that the chemical structure of an ionic molecule is not represented by its formula. Since there isn't a single ionic link between any particular pair of sodium and chloride ions, the term "molecule" for sodium chloride (NaCl) is erroneous. Ions are attracted to each other equally in all directions because attractive forces between them are isotropic or the same (Korchuganov et al., 2015). The ions then organize themselves into compact, three-dimensional lattice structures. For instance, sodium chloride is composed of an even ratio of Na^+ cations and Cl^- anions (Figure 5.8).

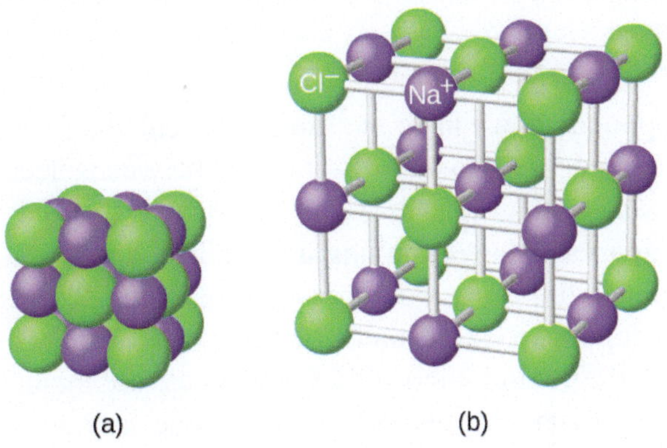

Figure 5.8. A sodium chloride atom in table salt are positioned to (a) increase the interaction of opposing charges. While the larger spheres stand in for chloride ions, smaller spheres represent sodium ions. The geometry is easier to see in the larger view. (b) Keep in mind that each surrounding ion—in this case, six—is "bonded" to every other ion.

Source: Inorganic Chemistry for Chemical Engineers.

In solid NaCl, the Na^+ and Cl^- ions are held together tightly by their strong electrostatic attraction. One mole of solid NaCl must be split into distinct gaseous Na^+ and Cl^- ions using 769 kJ of energy:

$$NaCl(s) \rightarrow Na^+(g) + Cl^-(g), \quad \Delta H = 769kJ$$

Did You Know?

Many important biological molecules are also inorganic compounds, equally important to that of enzymes, vitamins, and hormones – which are organic by nature. For example, the heme group in hemoglobin, which carries oxygen in the blood, contains iron atoms.

5.5. COVALENT BOND

Electrostatic forces cause ions with opposing charges to attract one another, and these ions are typically produced when metallic and nonmetallic atoms exchange electrons. Atoms connect in a certain way when attracted to one another due to a "shared" pair of electrons (Kitaigorodsky, 1978). These types of bonding are covalent bonds. Covalent bonds arise when two atoms have similar electron-attracting tendencies, or when their ionization energy and electron affinities are identical or nearly equal. For instance, when two hydrogen atoms join in a covalent bonding to form an H_2 molecule, each hydrogen atom is stabilized by two electrons, supplying it with the same valence electrons as a noble gas He.

Ionic compounds don't have the same physical characteristics as those with covalent bonds. Because the attraction between electrically neutral molecules is weaker than that between electrically charged ions, covalent compounds often have lower melting and boiling temperatures than ionic ones (Kadau et al., 2002). At room temperature, many covalent compounds are essentially liquids or gases, and when they do solidify, they frequently become much softer than ionic solids. Conversely, the majority of covalent compounds are electrically neutral and insoluble in water, making them subpar electrical conductors in both situations, while ionic compounds act as good electrical conductors when dissolved in water.

5.5.1. Formation of Covalent Bond

As mentioned in the early section, covalent bonds form through electron pair sharing of two or more like or unlike non-metallic substances. One classic example of covalent bonds that exist between 2 like non-metallic atoms, are the diatomic molecules, H_2, O_2, N_2, etc., while for those at 2 unlike non-metallic atoms is CH_4, CCl_4, etc. Figure 5.9 explains how this connection forms. We have two hydrogen atoms starting on the far right; the red line represents their potential energy. The separation between the two atoms is measured along x-axis. The two atom valence orbitals (1s) start to overlap as they get closer to one another (moving left along x-axis). As a result of their interactions with both atomic nuclei, the solitary electrons on each hydrogen atom fill space surrounding both atoms. The potential energy lowers as the bond distance shrinks because of the powerful attraction of each shared electron to both nuclei, stabilizing the system (Joly et al., 2006). As the atoms approach one another, the positive charges in the two nuclei begin to repel one another, raising the potential energy. The distance at which the lowest potential energy is attained determines the bond length (Janot et al., 1986).

The potential energy of the 2 hydrogen atoms decreases as they approach one another (right), and the lone electrons on each atom are shared to create a covalent bond, at which the bond length that yields the lowest potential energy is known as an *internuclear distance*.

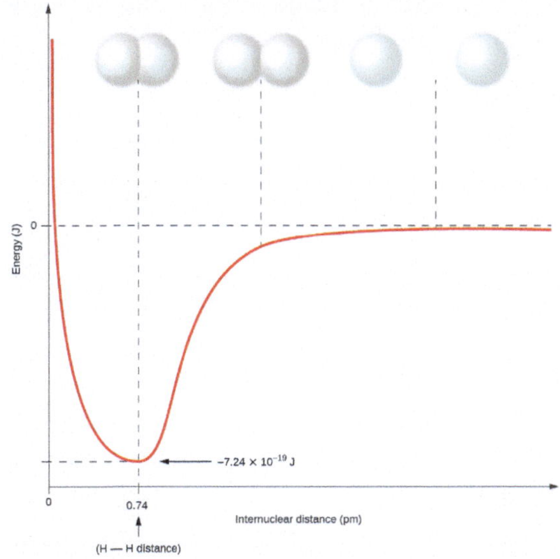

Figure 5.9. A potential energy diagram showing the interaction of two H atoms. The internuclear bond distance determines the lowest potential energy.

Source: Inorganic Chemistry for Chemical Engineers.

It is essential to remember that breaking chemical bonds involves adding energy (an endothermic reaction) while establishing chemical bonds releases energy (Jain, 2022). Because the covalent connection in H_2 is so strong, it takes a lot of energy—436 kJ—to separate atoms in one mole of hydrogen by rupturing or breaking the bonds between them:

$$H_2(g) \rightarrow 2H(g), \quad \Delta H = 436kJ$$

On the other hand, when two moles of H atoms unite, same amount of energy is released when one mole of H_2 molecules is formed:

$$2H(g) \rightarrow H_2(g), \quad \Delta H = -436kJ$$

5.5.2. Coordinate Covalent Bonding

Dative or coordinate covalent bonds are other bonds that develop between a Lewis base and an acid. Coordination is defined as donating a Lewis base to a Lewis acid without any extra bonding alterations. Lewis acid-base complexes are also called coordination complexes, particularly in transition metal chemistry (Holian & Lomdahl, 1998). A ligand is often used to refer to the Lewis base; in general, a ligand is only one molecule that binds to another (Figure 5.10).

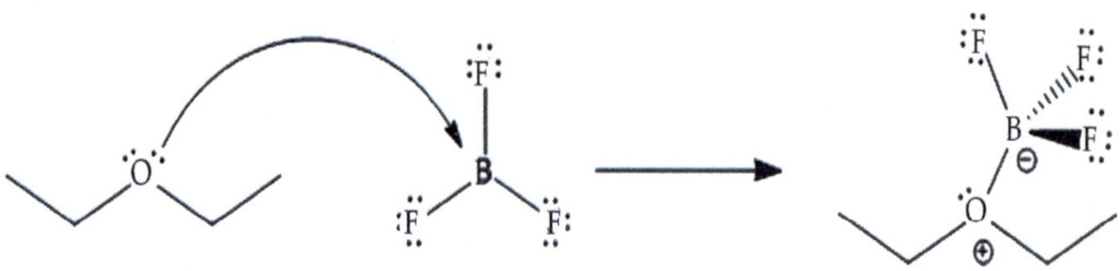

Figure 5.10. Lewis acid-base complex formation.

Source: Inorganic Chemistry for Chemical Engineers.

Hexaaquo cobalt dichloride, $Co(H_2O)_6Cl_2$, illustrates a coordination complex. A Co^{2+} ion can be found in this molecule (Gusev & Suter, 1993). Six nucleophilic water ligands coordinate this electrophilic metal ion. The compound has an overall charge equal to 2+ since the water molecules are neutral. To balance the charge, two chloride counterions are present, but they are not part of the ion complex (Figure 5.11).

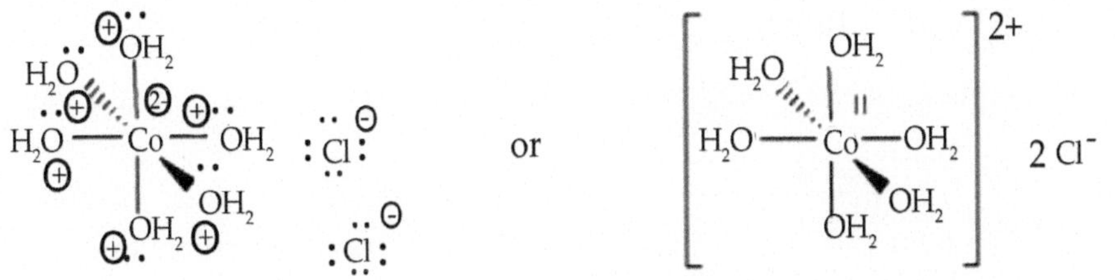

more common convention in coordination chemistry

Figure 5.11. Transition metal coordination complex representations.

Source: Inorganic Chemistry for Chemical Engineers.

Remember that when sketching coordination complexes of transition metals, the standard Lewis conventions are typically disregarded. Drawing the requisite formal charges becomes exceedingly challenging when so many distinct nucleophiles adhere to the Lewis acid. The metal cation's oxidation state is typically indicated (Gentile & Roy, 1960). The Roman numerals next to the metal atom denote the oxidation state, which is effectively charged on the metal cation. Also included is the overall charge for Lewis acid-base complex, which is shown by square brackets around charge. The reader must examine a place where the charge genuinely stands.

The mining and metallurgy fields commonly benefit from the utilization of coordination complexes (Geiger, 2016). For instance, nickel can be recovered from nickel ore by using carbon monoxide to transform the nickel into $Ni(CO)_4$. Tetracarbonyl nickel, often known as $Ni(CO)_4$, is a simple gas to separate from the solid ore. The coordination complex breaks down back into nickel and carbon monoxide when the of carbon monoxide is removed from the system (Figure 5.12).

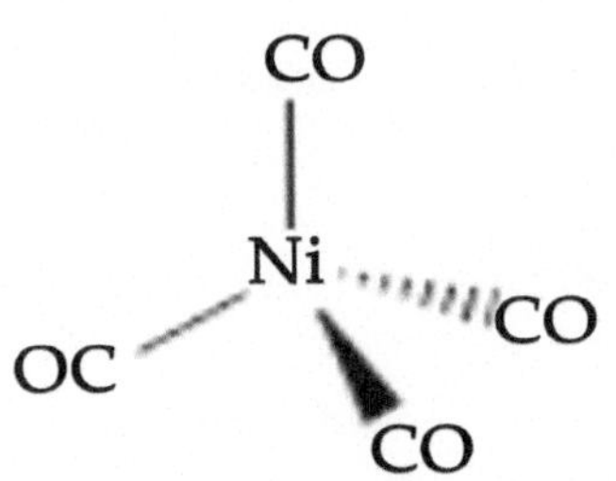

Figure 5.12. Tetracarbonyl nickel.

Source: Inorganic Chemistry for Chemical Engineers.

5.5.3. Metallic Bond

A "metallic bond" is collective sharing of a sea of valence electrons by multiple positively charged metal ions." The unique characteristics of metals such as their glossy sheen (lustrous), malleability, and their ability to conduct heat and electricity, result from this type of chemical bonding.

Some metal samples exhibit both metallic and covalent bonding (Fyfe, 1951). For instance, covalently bound gallium atoms frequently form metallic bound crystal formations. Additionally, the mercury ion displays covalent and metallic bonding.

The following elements can have an impact on a metallic bond's strength:

- The overall quantity of delocalized electrons;
- The positive charge of the metal;
- Ionic distance of the cation.

Example: One electron is located in the valence shell of sodium, which has an electron configuration $1s^2 2s^2 2p^6 3s^1$. Metallic sodium is an array of Na^+ ions surrounded by a sea of 3s electrons in the solid state. However, since all of the sodium cations share a sea of electrons, quenching the positive charge, it would be wrong to consider metallic sodium to be an ion (Figure 5.13).

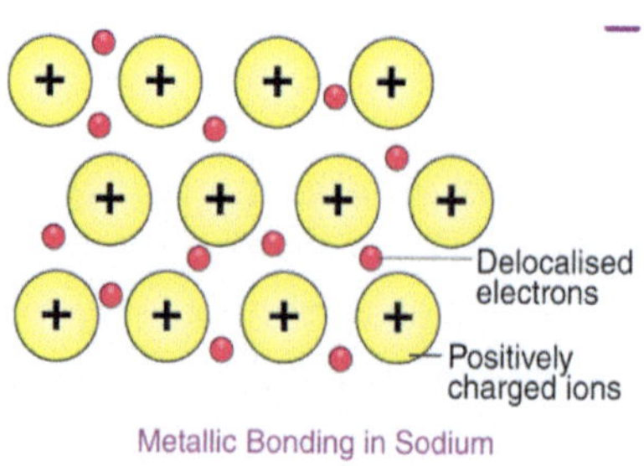

Figure 5.13. The metallic bonding in sodium.

Source: Byjus.com.

The low number of electrons in the electron sea and the low charge on sodium cation both contribute to sodium's softness and low melting point. For instance, a variety of Mg^{2+} ions make up metallic magnesium. Since two 3s electrons are delocalized into the sea, the electron sea in this system has twice as many electrons as that in metallic sodium (Fayzullin et al., 2017). Magnesium has a far higher melting point than sodium (650°C) because of the sea's higher electron density and greater magnitude of the charge.

Did You Know?

Since ancient times, inorganic materials have been utilized in art and architecture, such as lead-based paints for paintings and marble and limestone for construction materials.

5.6. PHYSICAL PROPERTIES OF INORGANIC COMPOUNDS

5.6.1. Density

The mass of the substance needed for that substance to occupy one unit volume is called its density. The specific gravity (relative density) or the ratio between a substance's density and a reference density, usually that of the density of water at 15°C, dimensionless, while. There many other units of measurement are used to express density, both in the SI and the traditional English units, including kg/m^3 and g/cm^3, and lb-mass/ft^3, respectively. Molar densities, computed by dividing the density by the molecular weight, are also frequently reported. Except as noted by the superscript figure, density values are presented at room temperature. For instance, the density of the substance at 15°C is represented by 2.487, which is equivalent to 2.487 g/cm^3. The density at 20° relative to water at 4° is shown by a superscript 20 above a subscript 4. The values are provided in grams per liter (g/L) for gases (Dergunova & Sakhnenko, 1994; Dirand et al., 2002).

5.6.2. Melting Point

The melting point of a solid is the temperature at which the vapor pressures of the solid and liquid phases are equal. The pressure is equal to one atmosphere, and the solid and liquid phases are in equilibrium. The freezing and melting points of pure substances are unique. The temperature at which a liquid solidifies under typical atmospheric pressure is known as the *freezing point*. On the other hand, the triple point is the point at which a substance's liquid, vapor, and solid phases exist in equilibrium. This is the point where the equilibrium lines for solid-vapor, solid-liquid, and liquid-vapor on a phase diagram converge. A phase diagram is a representation of a substance's state under various pressures and temperatures. The distinction is created in this way to show that the former has a melting point and decomposition at

250°C whilst the latter only experiences decomposition at 250°C and higher temperatures. Sometimes the melting point is given as 250 D, and in other times as d 250. When a number is given, such as $-6H_2O$, 150, it means that at 150°C, the chemical lost 6 moles of water for every kilogram of the formula. When hydrates melt chemically in water, that is the temperature at which they should be used (Dear, 1970).

5.6.3. Boiling Point

The temperature at which a substance vapor pressure matches that of surrounding atmosphere is referred to as *normal boiling point* (also known as boiling temperature). When a substance reaches its boiling point, it transforms from a liquid into a gas. The temperature at which a substance liquid and vapor (gas) phases can coexist in equilibrium accurately describes the boiling point. When a liquid is heated, its temperature increases until its vapor pressure is the same as the pressure of the surrounding air. Since the extra heat is being converted into latent heat of vaporization by the liquid, the temperature is not increasing rapidly anymore. This shift occurs both inside the liquid, where gas bubbles are created, and on its surface, where evaporation occurs. When pressure of the surrounding environment (gases) is decreased, the boiling point of the liquid decreases (Cruickshank, 1956).

On the other hand, the boiling point increases when the surrounding environment (gases) pressure increases. It's common to note that the pressure at which boiling point of a substance was determined when comparing it to the standard pressure of 760 mm of mercury or 1 atmosphere (STP, Standard Temperature and Pressure). One of the collative properties of all solutions is an elevated boiling point, which is typically greater than the pure solvent boiling point (Craig, 2007). Since the boiling point is usually written at atmospheric pressure (760 mm of mercury or 101 325 Pa), the notation 8,215 mm denotes the boiling point as 82°C. Subl 550 also asserts that the chemical sublimes at 550°C. Discussing breakdown products on occasion (Busing et al., 1964).

5.6.4. Refractive Index (RI)

The index of refraction or refractive index (RI), is determined by comparing the speed of light in a vacuum and that in a substance of a different density. The letters n or n' is frequently used in mathematical calculations and descriptive language to denote the RI variable.

In a vacuum, light values are expressed as a ratio of the substance air-saturated velocity to that of the air. Yellow sodium doublet lines, identified by the letter D and having a weighted mean of 589.26 nm, are the most prevalent sodium doublet lines. Using a mean value of 0.000 45 per degree for dn/dt and bearing in mind that nD drops as temperature rises, it is possible to calculate approximate values across a small temperature range when only one RI is available (Brock et al., 1987). Extrapolation is unreliable if a transition point is located inside the temperature range. Specific refraction, rD, is determined by the Lorentz and Lorenz equation,

$$r_D = n_D^2 - 1 / n_D^2 + 2\left(\frac{1}{p}\right)$$

The temperature-independent constant 'r' represents the RI and density. The molar refraction is obtained by multiplying specific refraction by molecule weight. It's an attribute that the various groups or components of the substance possess that is largely additive (Brock et al., 1982).

The empirical Eykman (McGraw Hill, 2003) equation:

$$n_D^2 - 1 / n_D^2 + 0.4\left(\frac{1}{p}\right) = constant$$

—enables the determination of the precision of experimental densities and refractive indices, as well as the calculation of one from the other, more precisely than Lorentz & Lorenz equation (1867). To get damp air RI, use the formula.

$$(n-1)\times 10^6 = \frac{103.49}{T}p_1 + \frac{177.4}{T}p_2 + \frac{86.26}{T}\left(1+\frac{5748}{T}\right)p_3$$

p_1 is the partial pressure of dry air (in mmHg), p_2 is the partial pressure of carbon dioxide (in mmHg), and p_3 is the partial pressure of water vapor (in mmHg), where T is temperature (in kelvins) (Brock et al., 1990).

5.7. APPLICATION OF INORGANIC COMPOUNDS

Inorganic compounds have a wide range of applications in coatings, catalysts, surfactants, pigments, medicines, fuels, and other fields. They are useful for particular uses because they often have high melting points and specialized high or low electrical conductivity qualities. For instance:

- In fertilizer, ammonia (NH_3) serves as a source of nitrogen, but also one of the main inorganic compounds used in the manufacture of explosives, fibers, nylon, plastics, polyurethanes, and hydrazine (used in rocket and jet fuels).

- Chemicals for water treatment and sterilization, agrochemicals (such as fertilizer, pesticide, or soil treatment), pharmaceuticals, and polyvinyl chloride (used for pipes, clothing, furniture, etc.) all use chlorine (Bredig, 1942).
- Naturally occurring titanium oxide, known as titanium dioxide, is utilized as a white powder pigment in various products, including coatings, plastics, paints, inks, textiles, paper, food, and cosmetics. Additionally, due to its strong resistance to ultraviolet radiation, it is increasingly needed in the syntheses of photocatalysts.
- Inorganic chemistry is also applied in:
- Both medical and healthcare profession.
- The usage of table salt or the chemical compound sodium chloride in daily life is the most widespread application.
- Cakes and other meals can be prepared using baking soda.
- The ceramics industry makes extensive use of inorganic chemicals (Bocharov et al., 2009).
- The field of electricity and electrical application specifically in electric circuits such as silicon in computers, etc.

Activity 5.1

Title: "Inorganic Compound Scavenger Hunt"

Objective: To explore and learn about the structure and properties of inorganic compounds, including chemical structures, ionic and covalent bonds, physical properties, and applications of inorganic chemistry.

5.8. SUMMARY

The structure and properties of inorganic compounds is a broad topic that covers various aspects of inorganic chemistry, including chemical structures, ionic and covalent bonding, physical properties, and applications.

Chemical structures in inorganic compounds can range from simple atomic structures, such as elemental metals, to complex molecular structures, such as coordination compounds or metal-organic frameworks. Understanding the chemical structure of inorganic compounds is critical for predicting their properties and reactivity. Ionic and covalent bonding are two fundamental types of bonding that occur in inorganic compounds. Positively and negatively charged ions are formed due to the transfer of electrons from one atom to another during ionic bonding. Metals, by transferring electrons to a non-metallic atom which then has the ability to accept the electrons, are the usual route to the formation of ionic compounds, while electron sharing between two non-metallic substance results in the formation of molecules or polyatomic ions through covalent bonding. Physical properties of inorganic compounds, such as melting point, boiling point, solubility, and density, are determined by their chemical structure and bonding. These properties are important for understanding the behavior of inorganic compounds under different conditions and for their practical applications. The applications of inorganic chemistry are vast and include areas such as materials science, catalysis, energy storage and conversion, environmental science, and medicine. Inorganic compounds are used in the production of semiconductors, ceramics, and metals for electronic devices, as catalysts for chemical reactions, and as drugs for treating diseases. Overall, comprehensive understanding of structure & properties of inorganic compounds is critical for advancing the field of inorganic chemistry and developing new and innovative applications for these compounds.

REVIEW QUESTIONS:

1. Describe the types of chemical structures commonly found in inorganic compounds, including simple ionic structures, layered structures, and complex structures such as zeolites.
2. Discuss in details the the applications of inorganic chemistry in various fields such as, (a) materials science, (b) environmental chemistry, and (c) medicine.
3. How do the properties of inorganic compounds relate to their applications in specific technologies, such as: (a)catalysis, (b) energy storage, and (c) electronic devices?
4. Explain the principles underlying the design and synthesis of new inorganic materials with specific properties, such as superconductivity or high strength.
5. Discuss the ethical considerations associated with the development and use of inorganic compounds in various applications, including their impact on human health, the environment, and natural resources.

REFERENCES

1. Abdelrahman, M., (2023). Imperfections in solids. *Biomaterials Journal, 2*(3), 1–16.
2. Abraham, N. L., & Probert, M. I., (2006). A periodic genetic algorithm with real-space representation for crystal structure and polymorph prediction. *Physical Review B, 73*(22), 224104.
3. Aswathappa, S., Arumugam, S., Sathiyadhas, S. J. D., Michael, J., Karupannan, K. B., & Amalapushpam, M. B. D. S., (2019). Modification of optical properties of ammonium dihydrogen phosphate crystal by employing shock waves. *Optical Engineering, 58*(10), 107101–107101.
4. Aswathappa, S., Palaniyasan, E., Dhas Sathiyadhas, S. J., Jayaperumal, K. S., Paramasivam, S., Sonachalam, A., & Dhas, S. A. M. B., (2020). Shock wave induced defect engineering on structural and optical properties of pure and dye doped potassium dihydrogen phosphate crystals. *Zeitschrift für Kristallographie-Crystalline Materials, 235*(6–7), 193–202.
5. Bartram, R. H., (1990). Radiationless transitions of point imperfections in solids. *Journal of Physics and Chemistry of Solids, 51*(7), 641–651.
6. Bekker, T., Rashchenko, S., Bakakin, V., Seryotkin, Y. V., Fedorov, P., Kokh, A., & Stonoga, S. Y., (2012). Phase formation in the BaB_2O_4–BaF_2–BaO system and new non-centrosymmetric solid-solution series $Ba_7(BO_3)_4$–xF2+3x. *CrystEngComm, 14*(20), 6910–6915.
7. Bocharov, S., Gille, P., & Glikin, A., (2009). Kinetic anomalies of mixed crystal growth and their effect on the crystal isomorphic composition. *Crystal Research and Technology: Journal of Experimental and Industrial Crystallography, 44*(1), 13–18.
8. Bredig, M., (1942). Isomorphism and allotrophy in compounds of the type A2XO4. *The Journal of Physical Chemistry, 46*(7), 747–764.
9. Brock, C. P., & Dunitz, J. D., (1982). Temperature dependence of thermal motion in crystalline naphthalene. *Acta Crystallographica Section B: Structural Crystallography and Crystal Chemistry, 38*(8), 2218–2228.
10. Brock, C. P., & Dunitz, J., (1990). Temperature dependence of thermal motion in crystalline anthracene. *Acta Crystallographica Section B: Structural Science, 46*(6), 795–806.
11. Brock, C. P., Minton, R. P., & Niedenzu, K., (1987). Structure and

thermal motion of triphenylboroxin. *Acta Crystallographica Section C: Crystal Structure Communications, 43*(9), 1775–1779.

12. Busing, W. R., & Levy, H. A., (1964). The effect of thermal motion on the estimation of bond lengths from diffraction measurements. *Acta Crystallographica, 17*(2), 142–146.
13. Craig, P. J., (2007). Ulrich muller. Inorganic structural chemistry Wiley – VCH, 2006, 268 pp; ISBN 13: 978□0□470□01864□4 (hardback)/10: 0□470□01864□X.
14. Cruickshank, D., (1956). The analysis of the anisotropic thermal motion of molecules in crystals. *Acta Crystallographica, 9*(9), 754–756.
15. Dear, P., (1970). Isomorphism of åkermanite and strontio-gehlenite. *Lithos, 3*(1), 13–16.
16. Dergunova, N. V., & Sakhnenko, V. P., (1994). Prediction of restricted isomorphism in solid solutions with the perovsktte structure. *Ferroelectrics, 154*(1), 59–64.
17. Dirand, M., Bouroukba, M., Chevallier, V., Petitjean, D., Behar, E., & Ruffier-Meray, V., (2002). Normal alkanes, multialkane synthetic model mixtures, and real petroleum waxes: Crystallographic structures, thermodynamic properties, and crystallization. *Journal of Chemical & Engineering Data, 47*(2), 115–143.
18. Eykman Formula, (2003). *McGraw-Hill Dictionary of Scientific and Technical Terms, 6E*. Retrieved from: https://encyclopedia2.thefreedictionary.com/Eykman+formula.
19. Fayzullin, R. R., Zakharychev, D. V., Gubaidullin, A. T., Antonovich, O. A., Krivolapov, D. B., Bredikhina, Z. A., & Bredikhin, A. A., (2017). Intricate phase behavior and crystal structure features of chiral para-methoxyphenyl glycerol ether forming continuous and partial solid solutions. *Crystal Growth & Design, 17*(1), 271–283.
20. Fyfe, W., (1951). Isomorphism and bond type. *American Mineralogist: Journal of Earth and Planetary Materials, 36*(7–8), 538–542.
21. Geiger, C. A., (2016). A tale of two garnets: The role of solid solution in the development toward a modern mineralogy. *American Mineralogist, 101*(8), 1735–1749.
22. Gentile, A., & Roy, R., (1960). Isomorphism and crystalline solubility in the garnet family. *American Mineralogist: Journal of Earth and Planetary Materials, 45*(5–6), 701–711.

23. Gusev, A. A., & Suter, U. W., (1993). Dynamics of small molecules in dense polymers subject to thermal motion. *The Journal of Chemical Physics, 99*(3), 2228–2234.
24. Holian, B. L., & Lomdahl, P. S., (1998). Plasticity induced by shock waves in nonequilibrium molecular-dynamics simulations. *Science, 280*(5372), 2085–2088.
25. Jain, V. K., (2022). *Defects in Solids Solid State Physics* (pp. 81–96): Springer.
26. Janot, C., Pannetier, J., Dubois, J.-M., & Fruchart, R., (1986). Isomorphism in quasiperiodic structures. *Physics Letters A, 119*(6), 309–312.
27. Joly, L., Ybert, C., & Bocquet, L., (2006). Probing the nanohydrodynamics at liquid-solid interfaces using thermal motion. *Physical Review Letters, 96*(4), 046101.
28. Kadau, K., Germann, T. C., Lomdahl, P. S., & Holian, B. L., (2002). Microscopic view of structural phase transitions induced by shock waves. *Science, 296*(5573), 1681–1684.
29. Kaiser, W., (1981). *Theorien der Elektrodynamik im 19. Jahrhundert* (pp. 157–162). Gerstenberg Verlag, Hildesheim.
30. Kitaigorodsky, A. I., (1978). Non-bonded interactions of atoms in organic crystals and molecules. *Chemical Society Reviews, 7*(1), 133–163.
31. Korchuganov, A. V., Zolnikov, K., Kryzhevich, D., Chernov, V., & Psakhie, S., (2015). Generation of shock waves in iron under irradiation. *Nuclear Instruments and Methods in Physics Research Section B: Beam Interactions with Materials and Atoms, 352*, 39–42.
32. Koshlyakova, N., Pekov, I., Zubkova, N., Agakhanov, A., Turchkova, A., Kartashov, P., & Pushcharovsky, D. Y., (2021). A new solid solution with garnet structure: Berzeliite–Schäferite isomorphic series from the fumarole exhalation of the Tolbachik Volcano, Kamchatka. *Geology of Ore Deposits, 63*(8), 857–868.
33. Kröger, F., & Vink, H., (1956). Relations between the concentrations of imperfections in crystalline solids *Solid State Physics* (Vol. 3, pp. 307–435): Elsevier.
34. Larson, A., Roof, R., & Cromer, D., (1964). The crystal structure of UF4. *Acta Crystallographica, 17*(5), 555–558.
35. Lennard-Jones, J. E., & Devonshire, A., (1939). Critical and co-

operative phenomena. III. A theory of melting and the structure of liquids. *Proceedings of the Royal Society of London. Series A. Mathematical and Physical Sciences, 169*(938), 317–338.

36. Li, S. S., & Li, S. S., (1993). Classification of solids and crystal structure. *Semiconductor Physical Electronics*, 1–19.
37. Lorenz, L., & Whittaker, E., (1867). *Phil. Mag: A History of the Theories of Aether and Electricity* (pp. 267–270). Nelson and Sons, London.
38. Luu, H.-T., Ravelo, R. J., Rudolph, M., Bringa, E. M., Germann, T. C., Rafaja, D., & Gunkelmann, N., (2020). Shock-induced plasticity in nanocrystalline iron: Large-scale molecular dynamics simulations. *Physical Review B, 102*(2), 020102.
39. Mestres, L., Martinez, M., Rodriguez, A., & Solans, X., (1987). The isomorphism between $CuSeO_4 \cdot 5H_2O$ and $CuSO_4$ $5H_2O$. *Zeitschrift für Kristallographie-Crystalline Materials, 180*(1–4), 179–188.
40. Murr, L., Niou, C., Jin, S., Tiefel, T., James, A., Sherwood, R., & Siegrist, T., (1989). Shock wave induced changes in superconductivity in $YBa_2Cu_3O_7$–δ. *Applied Physics Letters, 55*(15), 1575–1577.
41. Natta, G., Allegra, G., Bassi, I., Sianesi, D., Caporiccio, G., & Torti, E., (1965). Isomorphism phenomena in systems containing fluorinated polymers and in new fluorinated copolymers. *Journal of Polymer Science Part A: General Papers, 3*(12), 4263–4278.
42. Nelson, D. R., & Toner, J., (1981). Bond-orientational order, dislocation loops, and melting of solids and smectic-A liquid crystals. *Physical Review B, 24*(1), 363.
43. Nieminen, R., & Manninen, M., (1979). Positrons in imperfect solids: Theory. *Topics in Current Physics., 12*, 145–195.
44. Parker, M. C., & Walker, S. D., (2002). Arrayed waveguide gratings, fiber Bragg gratings, and photonic crystals: An isomorphic Fourier transform light propagation analysis. *IEEE Journal of Selected Topics in Quantum Electronics, 8*(6), 1158–1167.
45. Pekov, I. V., Zubkova, N. V., Agakhanov, A. A., Pushcharovsky, D. Y., Yapaskurt, V. O., Belakovskiy, D. I., & Britvin, S. N., (2018). Cryptochalcite, $K_2Cu_5O\ (SO_4)_5$, and cesiodymite, $CsKCu_5O(SO_4)_5$, two new isotypic minerals and the K–Cs isomorphism in this solid-solution series. *European Journal of Mineralogy, 30*(3), 593–607.
46. Poulain, A., Kubicki, M., & Lecomte, C., (2014). Isomorphism and

solid solution as shown by an accurate high-resolution diffraction experiment. *Acta Crystallographica Section B: Structural Science, Crystal Engineering and Materials, 70*(6), 973–982.
47. Puska, M., (1987). Theoretical aspects of positrons in imperfect solids. *Physica Status Solidi. A, Applied Research, 102*(1), 11–29.
48. Roth, J., (2000). Shock waves in quasicrystals. *Materials Science and Engineering: A, 294*, 753–756.
49. Seitz, F., (1958). Point imperfections in solids. *Il Nuovo Cimento (1955–1965), 7*, 414–443.
50. Shilkina, L., Talanov, M., Shevtsova, S., Kozakov, A., Dudkina, S., Nikol'skii, A., & Reznichenko, L., (2020). Isomorphism problems in lead-barium titanate. *Journal of Alloys and Compounds, 829*, 154589.
51. Showrilu, K., Jyothirmai, C., Sirisha, A., Sivakumar, A., Sahaya Jude Dhas, S., & Martin Britto Dhas, S., (2021). Shock wave-induced defect engineering on structural and optical properties of crown ether magnesium chloride potassium thiocyanate single crystal. *Journal of Materials Science: Materials in Electronics, 32*(3), 3903–3911.
52. Sivakumar, A., Saranraj, A., Dhas, S. S. J., Almansour, A. I., Kumar, R. S., Arumugam, N., & Dhas, S. M. B., (2021). Shock wave induced crystallographic structural phase transitions (Tri-states) of l-threonine crystal. *The Journal of Physical Chemistry C, 125*(45), 25217–25226.
53. Sørensen, A., & Mølmer, K., (1999). Quantum computation with ions in thermal motion. *Physical Review Letters, 82*(9), 1971.
54. Swaminathan, S., Craven, B., & McMullan, R., (1984). The crystal structure and molecular thermal motion of urea at 12, 60 and 123 K from neutron diffraction. *Acta Crystallographica Section B: Structural Science, 40*(3), 300–306.
55. Teitelbaum, S. W., Shin, T., Wolfson, J. W., Cheng, Y.-H., Porter, I. J., Kandyla, M., & Nelson, K. A., (2018). Real-time observation of a coherent lattice transformation into a high-symmetry phase. *Physical Review X, 8*(3), 031081.
56. Wang, F.-S., Yang, K., Kuo, Y.-R., Wang, C.-J., Sheen-Chen, S.-M., Huang, H.-C., & Chen, Y.-J., (2003). Temporal and spatial expression of bone morphogenetic proteins in extracorporeal shock wave-promoted healing of segmental defect. *Bone, 32*(4), 387–396.
57. Wherry, E. T., (1923). Volume isomorphism in the silicates. *American Mineralogist: Journal of Earth and Planetary Materials, 8*(1), 1–8.

58. Zvereva, I., Isaeva, A., & Choisnet, J., (2006). Cation distribution and interatomic interactions in oxides with heterovalent isomorphism: XI(La1–xHox)$_2$SrAl$_2$O$_7$ solid solutions. *Russian Journal of General Chemistry, 76*, 875–880.

CHAPTER 6

Structure and Optical Properties

Contents

Unit Introduction 146

6.1. Refractive Index (RI) 147

6.2. Refractive Indices: Effect of Composition, Structures, and Thermodynamic Conditions 152

6.3. Dipole Moments and Polarization 155

6.4. Szigeti Equation 157

6.5. Structural Application of Refractometry 160

6.6. Summary 163

Review Questions: 164

References 165

UNIT INTRODUCTION

Optical analysis techniques were restricted to a small band of wavelengths at close to the visible range roughly, a century ago. Now a days, this range includes radio frequencies as well as X-rays. In addition to the scope of the current book, an organized account of different techniques is not possible. Instead, we'll focus on using optical methods to solve problems that diffraction techniques can't solve, like those involving amorphous substances or brief species with half-lives on a scale of milli or even µscale (Grancini et al., 2014). Optical techniques allow for studying materials in various aggregate and thermodynamic states and provide significant insight into the workings of chemical bonding. Refractometry receives special attention in this section because its accomplishments and potential are currently underappreciated. In fact, refractometry has been used to study chemical structure and bonding since the middle of the 19th century (Sun, Chen, Sun, Guo, & Lu, 2003).

The reflection and structure turn out to have a more pronounced relationship than first thought. Refractometry was rendered obsolete by the spectacular advancements in diffraction techniques as well as NMR spectroscopy that occurred in the subsequent half of the 20th century. As a result, the initial, lofty expectations of turning it into a tool for determining structure were typically unfulfilled. This chapter will show that this method does indeed give unique, inaccessible knowledge regarding chemical bonds. Modern techniques can also directly measure polarization where classic refraction was able to make an addition to optical data (Zhu & Gao, 2014). Examples of these techniques include using atomic and molecular beams. A few remarks about one crystal's refractive indices before we wrap up. A century ago, it was a necessary skill for any chemist (let alone a crystallographer), but today it is all but lost, with the exception of geologists who use it to identify minerals in the field quickly. However, this technique can reappear, especially if it is reorganized using contemporary technology (Li, Hou, Pan, & Wang, 2009).

Remember

Light travels at a slower rate at higher refractive indices.

6.1. REFRACTIVE INDEX (RI)

6.1.1. Definitions, Anisotropy, and Theory

The changing electric field and (dynamic) polarization that occur if electromagnetic waves (which includes light) that pass through a substance have an impact on the wave speed. Because the speed of light (v) varies depending on the medium, a beam that crosses the interface among them gets refracted. The law of refraction, as Snelius and Descartes established in the 17th century, says that each pair of media has a refractive index (RI) that is constant for all incoming and refracted beam angles. The 19th century saw the discovery of the link between n and the proportion of light speeds in certain media through the equation (Singh, 2002):

$$n_{12} = \frac{\sin \theta_1}{\sin^0 2} = \frac{v_1}{v_v}$$

We calculate the *absolute* RI in this case if the speed of light *c* is constant in a vacuum (media):

$$n = \frac{c}{v}$$

where; RI is frequently utilized as a substance "fingerprint" in the laboratory. Since atmospheric air has a n = 1.00027≈1, measurements made in relation to it are typically accurate enough for use in all chemical processes (Ramakrishnan, Shrestha, & Soar, 2020).

All liquids, non-crystalline substances, gases, and cubic-shaped crystals are isotropic optically; i.e., n and v have same values in every direction, with a few significant exceptions. A wave formed by the surface of light coming from a single point of radiation in this medium will be spherical in shape. All other crystal optical characteristics rely on their crystallographic orientation. The "ordinary" beam, which has constant speed in each direction (v_o), and the "extraordinary" beam, which has variable speed in particular directions (v_e), are formed when a light beam penetrates a medium-symmetry crystalline (hexagonal, trigonal, or tetragonal) (Afromowitz, 1974). A rotation ellipsoid is created by the second beam while a spherical wave surface is created by the first. Only at 2 points, where the rotation axis exceeds the ellipsoid, do the two surfaces touch. This axis, which is also the major crystallographic axis, is known as the *optical axis*. Both conventional and striking beams move at the same speed in this direction. They are known as optically mono-axial crystals (Edlén, 1966). The optic indicatrix, which can be easily derived from the surface optics by changing overall radius-vector v along with its reverse, is a helpful method to explain the optical characteristics of an anisotropic crystal (Smith, Pendry, & Wiltshire, 2004) (Figures 6.1 and 6.2).

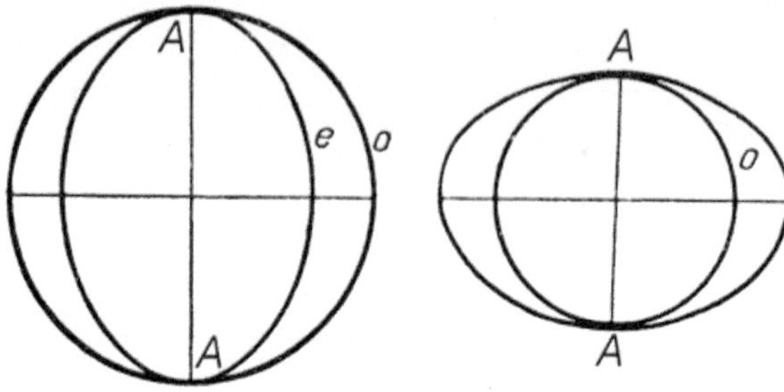

Figure 6.1. Optical surfaces of both the ordinary (o) and extraordinary (e) beams in mono-axial crystal and also the optically positive (left, $v_o > v_e$) and if the optically negative (right, $v_e > v_o$). AA is the optical axis.

Source: Wikipedia, Creative Commons License.

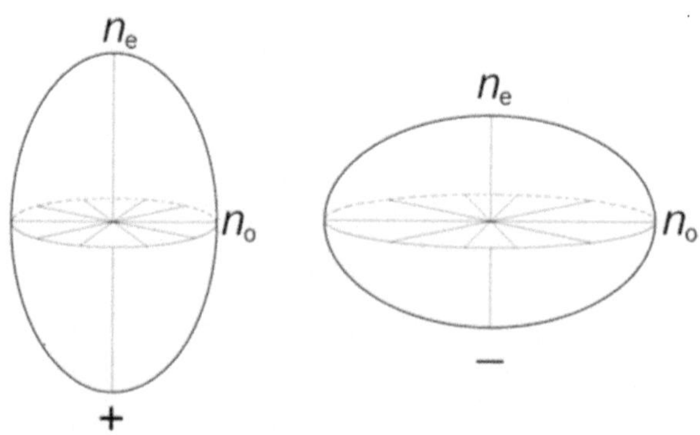

Figure 6.2. Illustration of optical indicatrices of mono-axial crystals: + for $n_e > n_o$, – for $n_o > n_e$.

Source: Wikipedia, Creative Commons License.

As a result, every property associated with surface optics is turned around. Additionally rotating on the principal axis of symmetry, the indicatrix is an ellipsoid that is prolate for optical optimistic crystal ($n_e > n_o$) and flat for optical negative crystal (no > ne). The optical anisotropy is measured by the distinction in n_o and n_e. Even in crystals with monoclinic, triclinic, and orthorhombic symmetry, a beam of light can split in two, but in these instances both beams are “extraordinary,” being polarized in the same plane. The letters n_g, n_m, and n_p (which stand for grand, moyen, as well as petite in French) are used to identify the three separate primary half-axes of the RI surface, which is an ellipsoid of generic kind (Kumar & Singh, 2010).

These ellipsoids are optically biaxial crystals because they have two planes (optical axes), whose opposite cross sections form a circle. Conversely angular crystal belongs to those with $n_g - n_m > n_m - n_p$, whereas optically $+v_e$ crystals have $n_g - n_m > n_m - n_p$. Be aware that optical mono-axial crystals may be classified as biaxial in nature having n_o = nm and $n_e = n_g/n_p$ and it depends on the optical sign of the crystal (bin Mat Yunus & bin Abdul Rahman, 1988).

Did You Know?

Anisotropy is a general feature of subsurface structure, and it can be caused by sedimentary layers sequencing or presence of subvertical fault systems.

A crucial implication is the capability of using measures of RI to differentiate among the three primary varieties of crystal symmetry. These include lower-orthogonal, monoclinic, or triclinic symmetry (crystals having 3 indices of refraction), medium, tetragonal shape, trigonal, and hexagonal symmetry, and high-cubic (optically isotropic crystals) (Fern & Onton, 1971).

Calculating the mean RI of an anisotropy crystal involves translating an ellipsoid into a circle with a similar volume.

$$\bar{n} = \left(n_e n_0^2\right)_{1/3} \qquad (6.1)$$

Dispersion is a property of refraction that depends on wavelength. Commonly referred to as nD, the sodium D line (589 nm or 2.10 eV) is used to determine RI for uniformity. According to Drude's theory, generally,

$$n^2 = 1 + \frac{N_1 e^2}{n^2 m} \Sigma \frac{C_i}{v_i - v^2} \qquad (6.2)$$

N_1 is the density of particle, e, and m are the electron's charge and mass, respectively, while c_i is the oscillator force, v_i is the sample absorption frequency, and is the measurement frequency. As a result, RI gets increased with the incident frequency toward an absorption band, where n =∞ at $v = v_i$, and is lowest at v = 0, or at indefinitely long wavelength (n_∞), and is reduced with increasing n in the i range (normal dispersion). It becomes higher with increasing n ∞ at $\nu > \nu_i$ (anomalous dispersion) (Veselago, Braginsky, Shklover, & Hafner, 2006). The spectroscopic method for measuring the unusual dispersion in crystals will be demonstrated below. Eqn. (6.3) takes the following form if there is only one main oscillator

$$n^2 = 1 + \frac{C_0}{E_0^2 - (hv)^2} \tag{6.3}$$

Here C_o is used as oscillator force while E_o is the single oscillator energy. h⁻ ν is the energy of photon. For ν = 0, Eqn. (6.4) adapts to a simpler type:

$$n^2 - 1 = \frac{C_0}{E_0^2} 4 \tag{6.4}$$

Wemple & DiDomenico (1971), utilizing the eminent Kramers–Kronigrelation, substituted C_o in Eqn. (6.3) by the product E_oE_d and here E_d is the dispersion energy, finding:

$$n^z - 1 = \frac{E_d E_0}{E_0^2 - (hd)^2} \tag{6.5}$$

They studied RI-dispersion data from over 100 different materials as well as liquids and found that a parameter called E_d, a calculation of the force of inter-band optic changes, follows an easy empirical formula.

$$E_d = \beta N_c N_c Z_a 6 \tag{6.6}$$

where; N_e is a practical amount of valence electrons, the formal charge of the anion is Z_a; N_c is the coordination number of the cation, and is a constant equal to 0.26 ± 0.04 eV for ionic compounds and 0.37 ± 0.05 eV for covalent substances (Foteinopoulou, Economou, & Soukoulis, 2003). The band gap (E_g) is approximately 1.5 times wider than E_d, so

$$(n^2 - 1) E_9 = \frac{N_c}{a} N_e Z_a \tag{6.7}$$

In other words, the ratio N_c/a indicates coordination number normalized, where a has a value of 4 for covalently compounds and 6 for ionic ones. The effective numbers of coordination N_c for these amorphous materials are derived by taking place of n and E_g of amorphous As_2S_3, Se, as well as Te into Eqn. (6.7). These values are in good accord with the results of X-ray diffraction. Shannon et al. (2002) utilized Eqn. (6.5) and an altered type of Sellmeier's equation to examine the RIs of 509 oxide and 55 fluorides.

$$\frac{1}{n^2-1}=A-\frac{B}{\lambda^2} \quad (6.8)$$

where; n was calculated for $\lambda = \infty$ (n_∞), where; λ is the wavelength. Low E_d as well as the comparatively large spreading of CuCl, TlCl, Na_2SbF_5 and NiF_2 were indicators of the impacts of formal valence, Z_a. Arsenates, vanadates, iodates, molybdates, Cu_2O, ZnO, and arsenites all showed the effects of cation coordination (Marple, 1964). The E_d of all hydrates is comparatively small. Due to the ambiguity in estimating N_e for some s^2 as well as d^{10} compounds, as well as when there are two or more cations that have distinct N_c, as in the case of $Y_3Fe_5O_{12}$, $CaTiO_3$, YVO_4, $CaMoO_4$ and more complex compounds like crystalline hydrates, it is not possible to determine E_d for the majority of compounds via the Wemple and DiDomenico scheme. Comparing E_d values based on the characteristics of anions like MoO_4^{2-}, BO_4^{5-}, BO_3^{3-}, PO_4^{3-}, SeO_4^{2-}, VO_4^{3-}, SO_4^{2-}, and WO_4^{2-}, the authors discovered that E_d is almost always influenced by the counter-ion (Vörös, 2004). It is challenging to determine E_d values in many multi-ion compounds because of these uncertainties and those relating to N_e and N_c. It is known that n empirically depends on E_g in a number of ways, viz.

$$n^4E_g := a\,,\quad n^2 = b - c\,InE_g, n = d - eE_g\,,\quad n^2E_g = f \quad \text{and} \quad n^z = 1+\frac{g}{E_g+h} \quad (6.9)$$

where; a, b, c, d, e, f, g, and h are constants. These formulas all result in identically high-quality results: RI increases when E_g decreases. This result seems obvious given that decreasing E_g raises the metallic properties of bonding as well as metals n. Although $n_{(ZnO)} = 1.94$ and $n_{(ZnS)} = 2.30$, respectively, and $E_{g(ZnS)} = 3.9$ and $E_{g(ZnO})) = 3.4$ eV, it is not always the case. Spectroscopy utilizing electron energy transfer tests were necessary to identify a substance's dielectric function $\varepsilon(\omega) = \varepsilon_1(\omega) + i\varepsilon_2(\omega)$, which may be used to determine the reflective index n as well as extinction coefficient k:

$$n_0 = 1+\frac{2}{\pi}\int_\infty^\infty \frac{k(\omega)}{\omega}\mathrm{d}\omega \quad (6.10)$$

The zero-frequency maximum of the real component $\varepsilon_1(\omega)$ is equivalent to n^2_o and is connected to the combined $\varepsilon_2(\omega)$ weighted by $1/\omega$

$$n_0^2 = \Sigma_1(0) = 1+\frac{2}{\pi}\int_\infty^\infty \frac{\varepsilon_2(\omega)}{\omega}\mathrm{d}\omega \quad (6.11)$$

It has been shown that the total absorption power per unit of volume of the seven allotropic phases of TiO_2, $TiOF_2$, andTiF_4, (anatasecolumbite, ramsdellite, brookite, rutile, hollandite, and bronze) provides a better explanation for their refractive indices than does their E_g.

$$I(\varepsilon_2) = \frac{2}{\pi} \int_{\infty}^{\infty} \varepsilon_i(\omega)\mathrm{d}\omega \tag{6.12}$$

As $I(\varepsilon_2)$ rises in the same order, the refractive indices of MX (M = Zn,Cd) gradually lowers down from MO<MS<MSe<MTe (Bond, 1965). This is caused by the reality that the covalent bonding strength in the ZnX and CdX bonds increases in the following order: O<S<Se<Te.

6.2. REFRACTIVE INDICES: EFFECT OF COMPOSITION, STRUCTURES, AND THERMODYNAMIC CONDITIONS

These dependencies can be demonstrated by contrasting various polymorphs. In glass and films, when diffraction methods are not particularly useful, one can estimate N_c because it is known that RI rises alongside N_c. Glass is created when crystalline GeO_2 is compressed violently. The heterogeneous result contains both grains with n = 1.8-2.0 and those with n = 1.608–1.610, indicating a rise in Ge's N_c to 6, that was later established in studies employing static compression on vitreous GeO_2. It is widely recognized that the earliest layered crystalline substances placed on substrates with various crystal structures either imitate the latter or undergo a phase transition. High-grade crystals are necessary for optical anisotropy measurements (Pan & Trusler, 2022).

However, symmetry becomes increasingly muddled and the specimen can turn out to be pseudo-isotropic as the concentration of imperfections rises and/or the size of the crystal shrinks. After shock-wave treatment, crystals show this the most clearly. As a result, shocked silicates showed RI values that were both significantly lower and less anisotropic. The stronger the effect, the greater the pressure (Reis, Lampreia, Santos, Moita, & Douhéret, 2010).

Did You Know?

Thermodynamics gives the framework for several other crucial ideas that the modern world depend on, including but not limited to nuclear power, heat engines, freezers, chemical reactions, and others.

The optical anisotropy was restored after annealing the defects in the shocked crystals, which contained firmly disordered grains and even amorphous regions. Shock-treated quartz samples revealed an immediate drop in the density, RI, and anisotropy due to the formation of vitreous patches, whereas remaining crystalline areas revealed an extended unit cell (Maligne & Bruneaux, 2011). The density (through as much as 11%) and RI of quartz glass were later shown to rise with stress compression up to the value of P = 26 GPa but diminish with more severe shock waves. N_c and d(Si-O) of the condensed glass were unchanged, and the density increased as a result of more closely packed SiO_4 tetrahedral, according to XRD examination. The formation of crystallographic defects, including electronic ones, decreased the density and RI in the shocked polycrystalline LnF_3. Real-time studies of the nature of chemical bonds are possible, and thanks to measurements of the RI of crystals through heating (Notomi & Mitsugi, 2006).

As a result, ionic and semiconductor crystals have different values for the factor $\delta = 1/n\ \partial n/\partial t$. In the latter, RI only decreases as a result of reduced density (thermal expansion), while semiconductors simultaneously experience an increase in polarizability as a result of the appearance of free electrons. It was possible to group materials based on their $\boldsymbol{\delta}$ when considering the expansion and electronic factors. RI can reveal the structural function of a material's various components if it alters its composition when heated (Wanstall, Agrawal, & Bittle, 2020). Dehydrating zeolites results in a decrease in RI because the density is decreased but the framework basically stays similar, whereas dehydrating typical crystalline hydrates boosts their RI as the structure changes into a denser one. The RI increases monotonically when hydrogen as well as water condenses at pressures up to 35 GPa, and the polarizability of these solids does not significantly change when they fuse. Ionic and semiconductor crystal RI also rises with density, with the exception of MgO and Al_2O_3, which only differ in this regard due to their different compressibilities; a reduction in RI beside the c axis was discovered following a shock compression (Tariq, Forte, Gomes, Lopes, & Rebelo, 2009).

The largest increase in RI was seen in CsH, going from 1.28 at atmospheric pressure to 3.20 at P = 253 GPa. Under P≤130 GPa, the RI of H_2 changes to n = 0.687343 + 0.00407826P + 1.86605(0.29605 + P)$^{0.0646222}$, while that of BeH_2 changes to n = 1.474 + 0.0868P$^{0.00245P2}$. Between 208 and 288 GPa, the RI of CH_4 exhibits a sharp increase, pointing to a change in phase of the insulator to semiconductor type. The SiH_4 RI alters between 109 and 210 GPa as n = 0.33955 + 0.02332P, that also denotes an insulator to semiconductor change at P = 109 GPa, but between 7 and 109 GPa as n = 1.5089 + 0.00349104P (Barth, Goncalves, Pires, Roeder, & Wolf, 2000) (Figure 6.3).

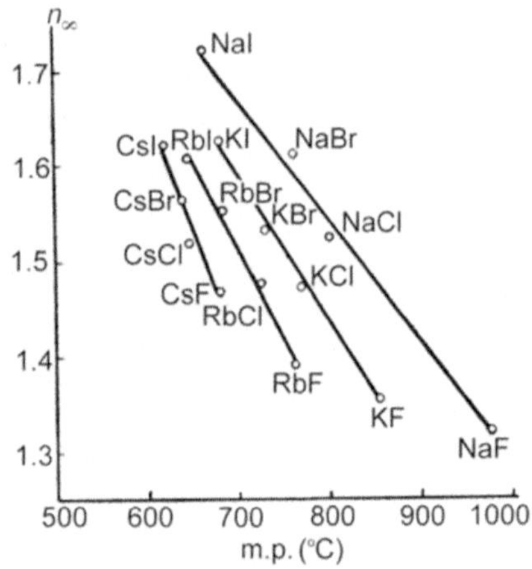

Figure 6.3. Refractive indices (n∞) of crystalline alkali halides depends on the melting temperature.

Source: Wikipedia, Creative Commons License.

RI has significantly increased. Over 0.6 GPa, the RI of CO_2 differs because n = 1.41P$^{0.041}$. According to Eqn. (6.3), for compounds with the similar structure, RI becomes higher as the bonding becomes more covalent and metallic. As the absorbing frequency (vi) decreases depending on the conversion from ionic to covalent to metallic compounds, the value of the denominator of Eqn. (6.3) approaches zero, resulting in n. This circumstance will be used later to calculate atomic polarizabilities and forecast the levels of pressure where fundamental solids and compounds will transform into metallic. RI relates to the molecular-physical characteristics of the material. As a result, the RIs of crystalline alkali halides linearly decline with increasing melting temperatures, while RIs for organic liquids show a relationship with boiling temperatures (Malham & Turmine, 2008).

6.3. DIPOLE MOMENTS AND POLARIZATION

The evident region of the spectrum is used to calculate the RI. Hence, the dielectric constant (ε), which is obtained when these measurements are made in the infrared (IR) spectrum or with even longer waves, is distinguishing. Maxwell's law states that $\varepsilon = n^2$ for covalent substances and $\varepsilon > n^2$ for polar compounds. The difference results from two polarization mechanisms: electronic polarization (P_e), which only affects the electronic clouds of atoms and molecules, and atomic polarization (P_a), which affects all atoms with net charges (Liptay, 1974). According to Debye's theory, molecular polarization (P_m) is typically represented as

$$P_m = v\frac{\varepsilon - 1}{\varepsilon + 2} = \frac{4}{3}\pi^N\left(a + \frac{u^2}{3KT}\right) \tag{6.13}$$

where; V denotes the volume of molecule; N denotes Avogadro's number, the polarizability of molecule, and its dipole moment. When $\varepsilon = n^2$ for non-polar substances, Eqn. (6.13) takes the shape

$$P_e = V\frac{n^2 - 1}{n^2 + 2} = \frac{4}{3}\pi Na \tag{6.14}$$

The equation by Lorentz-Lorenz is so called the polarizability and the dipole moment may be determined independently by measuring P_m at various temperatures. Spectroscopic techniques can also be used to measure the dipole moments of gases. Dipole moment information is of autonomous interest, and there is a sizable body of literature on the subject. Recent measurements of metallo-fullerenes, M • C60, revealed that their dipole moments ranged from μ=12.4 D for M = Li to 16.3 D for M = Na to 21.5 D for K to 20.6 D for Rb to 21.5 D for Cs (Hickey & Rowley, 2014). When defining the dipole moment, in the initial approximation:

$$\mu = e^* d \tag{6.15}$$

The length of dipole d corresponds to the distance of bond and e* is equivalent to the actual charges of the atoms. Pauling later utilized this restriction to ascertain the ionicity of numerous chemical bonds. Fajans used this ratio to determine the bond polarities in HBr and HCl. As a result of charges compensating for each other, μ of NO is less (0.16 D) and of CO is almost zero. Eqn. (6.15) is therefore a very basic simplification (Amos & Burrows, 1973). Coulson concluded that the dipole moment of a bond is made up of numerous terms, including the difference in electronegativity

(μ_i), the difference in density of electron (μ_p), the hybridization of the atomic orbitals (μ_h), as well as the impact of non-bonding electrons (μ_e):

$$\mu = \mu_i + \mu_p + \mu_e \quad (6.16)$$

Therefore, simply put, if other elements of μ were referred to hypothetically or empirically could the bond ionicity be determined from the determined dipole moments. To get around this issue, we can contrast a number of related compounds where regards other than i are either missing or make up for each other. Pauling used this method to explain how the value of μ/d depends on the dissimilarity in electronegativities using the molecules of HX as well as alkali halides (Targema, Obi-Egbedi, & Adeoye, 2013). The dipole moment in a polyatomic molecule is equal to the sum of all bond moments, so, μ = 0 in a symmetrical molecule (like MX_4) but μ = 0 in an asymmetrical one. μ ≤ 0.2 D in $M(C_6H_6)_2$ complexes with M = Sc, Ti, V, Nb, Ta, while $Co(C_6H_6)_2$ and $Ni(C_6H_6)_2$ have μ = 0.7 and 1.3 D, correspondingly (Jasien & Fitzgerald, 1990).

Surprisingly, complexes in the gas phase made of neutral atoms as well as molecules can contain very large dipole moments. For instance, the complex $X(C_6H_6)$ with X = Cl, Br, and I has μ = 1.712, 1.715, and 1.625 D, correspondingly. The measured dipole moments of combined rare gas dimers are μ (N_e • Ar) = 0.003, μ (Ne • Kr) = 0.011, and μ (Ne • Xe) = 0.012 D; these indicate negligible bond polarity at the respective interatomic distances of 3.49, 3.63, and 3.86 Å (Jensen, Åstrand, & Mikkelsen, 2001).

In spite of the qualitative validity of these contrasts, the accurate relationship among polarization as well as the electronic makeup of a substance has only been discovered using the spectroscopic method. This method has allowed scientists to independently find out the atomic (P_a) as well as electronic (P_e) polarizations of atoms as well as their effective charges by analyzing the magnitudes of IR absorption bands in molecules (Karlström & Sadlej, 1982).

It is obvious that the polarity of a bond must be qualitatively described by the ratio Pa/Pe. The spectroscopic measurements of atomic sets because anharmonic and variations in orbital overlaps during atomic vibrations are ignored. Using the polarization idea for solids, together with ionic compounds, is crucial. Utilizing experimental data for and V and Eqn. (6.13), the additive technique was used to compute the molar polarization of 129 oxides and 25 fluorides of a variety of metals to predict the polarization of 61 ions. Using tabulated ionic polarization values to analyze the molar polarization of numerous different ionic crystals enables one to infer the presence of particular characteristics such as cation deformations, piezo- or ferroelectricity or deformations in bond character based on the deviation from the average value. The mechanism of crystal polarization was considered by Born's theory, which proposed that within an inorganic crystalline material, atomic interactions produce ionic attraction and electronic repulsions (Pitoňák, Holka, Neogrády, & Urban, 2006). Born related ε the electric fields and n to the incidence of the transverse lattice vibrations (ω_t), considering that the electric field performing on every ion is equivalent to the external field applied:

$$\varepsilon = n^2 + \left(\frac{4\pi e^2}{\omega_t^2 \overline{m} v} \right) \quad (6.17)$$

Where the other symbols are as described above, and m is the decreased mass of the vibrating particle. Comparison of theoretical values of (calculated using the born formula) and experimentally measured values of revealed that, typically, the theoretical values were lower. Ions are not fixed spheres, and P_a and P_e are not additive, which results in this discrepancy. Actually, the atomic and electronic polarizations are no longer separate additive quantities because the electron clouds of the ions in the crystals have misshapen as well as merged at the lattice vibrations. Szigeti (1949) demonstrated that the atomic polarization is increased by the distortions of the electron cloud of ions brought on by the lattice's transverse vibrations (Liptay et al., 1988). Szigeti arrived at the following equation by considering that the polarization of the sample causes the internal field force in the dielectric to be lower as compared to the external field:

$$\varepsilon = n^2 + \frac{4\pi e^2}{\omega^2 \overline{m} v} \frac{n^2 + 2}{9} \quad (6.18)$$

6.4. SZIGETI EQUATION

The electronic polarization's contribution is considered in the Szigeti equation by the factor $(n^2 + 2)^2/9$, which makes it different from the Born formula. This term is more important since, the higher the RI, or covalency of the bond, is. The values determined by Eqn. (6.18) are forever greater as compared to those determined through experimental measurement. This gave Szigeti the idea to replace the ion's formal charge with a parameter s that describes how the actual charge, e*, differs from its theoretical value

(Baonza, Taravillo, Caceres, & Nunez, 2006). It is clear that the s parameter in a physical sense equals the atom's effective charge: s = e*. Szigeti (1949) discovered that when $\omega_t = 2\pi c v_t$, where *c* is the speed of light and t is the frequency of the lattice's transverse vibration

$$e^* = \frac{3v_t}{Z_e\left(n^2+2\right)}\left[\pi\left(\varepsilon - n^2\right)\bar{m}v^1\right]^{\frac{1}{2}} \tag{6.19}$$

Here; Z is the atom's valence. The Szigeti equation was first applied to determine efficient charges of atoms in straightforward cubic crystals and afterward, extra intricate crystalline compounds, both anisotropic and isotropic, were investigated. The growth of the Szigeti method is correlated with advancements in measurement methodology and the method for computing the optical property of anisotropic crystals, y. Because the IR-absorption band shifts monotonically in every spectral interval from v_l to t vibrations, transverse vibration frequencies can only be found out on single crystals; determining them on powdery substances is very difficult (Everitt et al., 2017) (Figure 6.4).

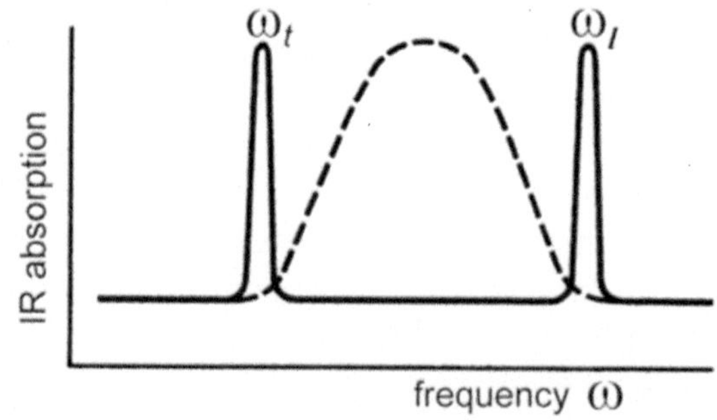

Figure 6.4. *IR absorption of a single crystal shown in (solid line) and powder shown in (dotted line).*

Source: Wikipedia, Creative Commons License.

However, the equation by Lyddane, Sachs, & Teller (1941), which links the greatest amount of IR absorption with the ω_l and ω_t frequencies of the lattice vibrations, is of realistic implication,

$$\left(\frac{\omega_t}{\omega_t}\right)^2 = \frac{\varepsilon}{n^2} \tag{6.20}$$

According to this formula, if $\varepsilon = n^2$ the transverse and longitudinal frequencies are equal, meaning that the IR absorption band width is zero.

Actually, according to spectroscopic theory, the magnitude of IR-absorption is zero not only when the atomic vibrations in question have no impact on the dipole moment except when μ = 0, or in the matter of solely covalent materials where $\varepsilon = n^2$. An alternative strategy is predicated on the idea that all of the outer electrons of atoms in a crystal can be delocalized, just like in metals, and that the laws of an electronic gas apply to them. By keeping with Penn's theory

$$n^2 = 1 + \left(\frac{h_{\omega\pi}}{E_g} \right)^2 \tag{6.21}$$

where; ω_p is the plasma frequency of the valence electron vibrations and E_g is the width of band gap.

$$\omega_p = \left(\frac{4\pi e^2 N}{mV} \right)^{1/2} \tag{6.22}$$

Here; N denotes the total valence electrons, m for the mass, and V is the crystal's formula unit volume. Phillips and van Vechten (1969) established the *dielectric theory of a chemical bond* based on these relationships. Due to the fact that silicon and carbon (which make up diamond) have the formulas $\varepsilon = n^2$ and $E_g = E_h$, respectively, the equation requires the form:

$$\varepsilon = 1 + \left(\frac{h\omega p}{En} \right)^2 \tag{6.23}$$

Phillips derived covalent (E_h) and ionic (C) Coulomb from Eqn. (6.23), assuming that it is applicable to all crystal types.

$$E_g^2 = E_h^1 + C^2 \tag{6.24}$$

Eqns. (6.21) and (6.23) can be calculated from Eqn. (6.24) and shows substantial agreement with the experiment (Szigeti, 1949). The relationship between C and atoms charges and radii can be written as,

$$C = a \left(\frac{2A}{r_A} - \frac{ZB}{r_B} \right) f \tag{6.25}$$

The valence is Z, the radius is r, and the Thomas-Fermi screening factor is f. The second shows thus completely applicable to Thomas-Fermi statistics. Phillips calculated the atomic electronegativity as using Eqn. (6.25) as:

$$X = 4.0\left(\frac{z}{r}\right)f + 0.5 \tag{6.26}$$

Significant E_h as well as C, we may conclude the bond ionicity as

$$f_j = \left(\frac{C}{E_g}\right)^2 \tag{6.27}$$

It is now challenging to list all the studies that use or develop the dielectric theory of chemical bonding since the Phillips and van Vechten theory is a crucial tool for solid state chemistry as well as physics. Only the basics of this strategy have been discussed here; the next section will cover some additional elements and uses (Hush, Szigeti, Carvalho, & Hope, 2013).

6.5. STRUCTURAL APPLICATION OF REFRACTOMETRY

Refractometry was once used to determine the chemical structures of silicates, geometrical isomers in compounds with coordination, structural formulae for organic chemistry, and other things. However, it is still useful for reviewing hydrogen bonds and the interaction of atoms. The H...Y parting becomes less but the XH separation rises due to the development of a hydrogen bond (H-bond) in the X-H...Y system. However, as what was previously demonstrated, the other crystal chemical variables also impact the distances, making it difficult to distinguish between various effects. Therefore, developing new techniques for investigating H-bonds is a crucial task (Dorrah, Zamboni-Rached, & Mojahedi, 2018). Comparing the molecular refraction of substances containing and absence of H-bonds can provide useful information. If the anion X is a worthy hydrogen bond acceptor, the difference $\Delta R = R(NH_4X)\ R(KX)$ is 1.5 cm^3/mol in the lack of H-bonds, but it gets increased by 0.2 to 1.0 cm^3; a similar contrast of $MX_n \bullet mH_2O$ and MX_n yields $R(H_2O)$ = 3.5 to 3.9 cm^3/mol for a good acceptor X and 3.40 cm^3/mol for a poor one. Interestingly, when $K_2M(CN)_4 \bullet 3H_2O$ and its dehydration product were compared, $R(H_2O)$ was even lower, coming in at 2.0 cm^3/mol for M = Pt and 2.4 cm^3/mol, for M = Pd (Surdo, Carpignano, Strambini, Merlo, & Barillaro, 2014).

Due to the formation of MNC bonds during dehydration, this effect turned out to be a hoax; accounting for them resulted in reasonable $R(H_2O)$

values of 3.5 as well as 3.9 cm³/mol, accordingly, Refractometric data led to a hunch about the rearrangement, which was later verified by NMR- and IR-spectroscopy. H-bonding is indicated by the liquid water's refraction, which is 3.66 cm³/mol. Water's RI increases under shock compression from 1.33 at ambient pressure to 1.60 at P = 22 GPa. Metallization was once used to explain this increase, accompanied by a sharp increase in electric conductivity. The $R(H_2O)$ = 3.2 cm³/mol, which is in fact less than medium and denotes the breakage of H-bonds, is simply due to the high density if compressed water. Proton mobility is increased as a result, increasing conductivity. The "metallic water" theory was directly refuted by refractometry, which confirmed this conductivity mechanism (Rong et al., 2012).

The polarity of the nitrogen and oxygen and particles is raised in acid salts and acids by the buildup of hydrogen atoms in the outermost region of a complex ion, strengthening the H-bonds with those atoms. This pattern is caused by variations among the relative charges of the atoms. The O– and N–H-type bonds with the identical anions exhibit analogous refraction due to the similar operating charge of the hydrogen molecules in H_2O and NH_4^+ (Chen, Liu, Yang, Wilkinson, & Zhou, 2019).

Because of various alterations on the oxygen atoms in such compounds, the refractions of H-bonds are approximately 0.2 cm³/mol in aliphatic alcohols and between 0.2 and 0.4 cm³/mol in aromatic alcohols. The H-bond refraction of alcohols is improved when electrophilic substituents like OH^- or Cl^- are added, that is in line with the fact that adding electrophilic groups raise the boiling points of alcohols. The H-bond refraction becomes less intense with longer carbon chains in aliphatic acids. It is important to emphasize that the relationship of the H-bond energy and the atomization energy only accounts for a small percentage of the intrinsic refractions of substances. Because the refraction of the oxide constituent of silicates is dependent on N_c as a cation, refractometric approaches are helpful for studying vitreous or polycrystalline silicates to ascertain the chemical function of anions or to describe N_c of cations (Schroeder, Ecke, Mueller, Willsch, & Andreev, 2001).

$$R_N = (R_1 N_i)^{-1/3} \tag{6.28}$$

where; $N_c = 1$ and R_1 represent the similar oxide refraction in molecular type. To investigate the trans-effect in coordination compounds, the refractive method is particularly helpful. This increases in succession in terms of Pt(II) complexes

CO, CN, C_2H_4> NO_2, I, SCN > Br, Cl> OH, F > NH_3, H_2O

This shows that the rise in bond polarity will make these ligands more soluble in aqueous solutions, and must be related to the negative operational charges on the ligands affected by the trans-partners. The optical method revealed the first quantitative property of the trans-effect: the refraction of the bond electrons, which is the distinction among the atomic and ionic refractions of ligands and platinum, variations in the series correlating to the trans-effect sequence: CO_3> NO_3, SO_4>F > NH_3>H_2O, H > S > SCN > I > Br > CN >Cl> NCS > NC > NO_2> OH, and NCS > NC > NO_2> NCS (Skafte-Pedersen, Nunes, Xiao, & Mortensen, 2009). The atom bonded to Pt allows us to group these radicals neatly:

F < O (ONO, SO_4, NO_3, CO_3, OH) < N (NO_2, NCS, NC) < S (SCN, S)

The SCN, CN, and NO_2 categories have significantly different functions depending on what part of the ligand links to the central atom, which is the fundamentally novel feature of the preceding sequence. These radicals can be connected in various ways in complex compounds of a variety of metals, as well as in complex compounds of a single metal, and in such circumstances, the order of trans-activity will be quite distinct. We can quantify the trans-effect if we compare the coordinate refractions (Barer, 1957).

This means that $\Delta R = R(X\text{-}M\text{-}Y) - 1/2\ [R(X\text{-}M\text{-}X) + R(Y\text{-}M\text{-}Y)]$ is a representation of the trans-activity of the X and Y atoms. The technique has shown that a NO_2 ligand possesses a larger trans-effect than Cl^- in coordination compounds of Pt(II), but the opposite is true for Pt(IV). The complicated Co(III) and Pd(II) compounds, which similarly defied the paradigm at that point but were later established by different physical approaches, also led to similar conclusions. The mobility coefficient (MC) is the ratio of the basic fundamental refraction of the complex molecule in question to the atomic refraction of an impacted ligand. The computations of MC is described (Batsanov, 1961). Depending on how responsive they are to the trans-effect, they deliver the ligands in the order indicated below,

H»NH_3> H_2O >Cl> NO_2> SCN > Br

This meand that no other physical approach could characterize similarly, but which has been verified by chemical experiments. The calculation of atomic sizes is a crucial component of structural refractometry. Wasastjerna (1939) was the first to calculate ionic radii using the formula = r^3, which the Clausius-Mossotti theory requires (Wang et al., 2011). Goldschmidt utilized the Wasastjerna-calculated fluorine and oxygen ionic radii to derive other radii using the additive method. Pauling's univalent ionic radius and refraction have an empirical relationship, according to Kordes' research, which he began in 1939,

$$R \cdot i = k_r \cdot i^{4.5} \tag{6.29}$$

where; k equals 0.603 for rare-gas ions and 1.357 for all other ions. Kordes (1940) discovered the refractions of free ions via these ionic radii, and subsequently estimated crystalline ionic refractions, presuming r_1 to be typical of a free ion. Wilson, Curtis, & Vieillard (2011) found a correlation among the refractions and radii of the independent and crystalline ions. Alkali fluorides that have been melted have shown a similar dependence (Wu, Semenova, Wang, & Farrell, 2011). Miller described the interdependence between the van der Waals radii (rW) and the polarizabilities of atoms that are isolated because in accordance with the Slater-Kirkwood theory

$$a = \frac{4}{a_0}\left(\frac{r_w^2}{3}\right)^2 \tag{6.30}$$

Then, using the atomic refractions, the vdW radii can be calculated, that deviate by 10–30% from the standard values. Fullerenes and its derivatives somehow showed some intriguing results, with $R\infty(C_{60}) = 193$ and $R\infty(C_{70}) = 257$ cm^3/mol, respectively, according to the additive model. Formally, C_{60} consists of 60 single bonds and 30 double bonds, which results in $R(C_{60}) =$ 193.4 cm3/mol for the typical bond refractions of 1.254 (CC) and 3.94 cm3/mol (C=C). The refraction of sodium adducts of fullerene is much higher than that of pristine material or atomic sodium (60.2 cm^3). The refraction of sodium adducts of fullerene decreases sequentially as $Na{\bullet}C_{60}$ 4410, $Na_2{\bullet}C_{60}$ 2520, and $Na_3{\bullet}C_{60}$ 1,890 cm^3/mol, and then increases slowly with the digit of sodium atoms to arrive at 5,040 cm^3/mol for $Na_{34}{\bullet}C_{60}$. This effect was accounted for by forming a permanent dipole and a charge transfer from Na to fullerene (Banerjee, 2009).

6.6. SUMMARY

Anisotropy, optical, and structural characteristics are outlined in this chapter. Optically isotropic materials include gases, cubic symmetry crystals, liquids, and non-crystalline solids (with a few notable exceptions). These dependencies were highlighted by contrasting various polymorphs and the effect of composition, thermodynamic situations, and structures, on refractive indices, are fundamentally explained. The optical anisotropy returned after heating the defects in the shocked crystals, which contained firmly disordered grains as well as amorphous regions. Due to the development of vitreous areas, shock-treated quartz samples displayed an immediate reduction in density, RI, as well as anisotropy, at the same time as the remaining crystalline areas displayed prolonged unit cell. This chapter has described the need to point out that H-bond refractions include only a small percentage of the inherent refractions of substances, which determine the chemical structures of silicates, the structural formulae of organic compounds, geometrical isomers in compounds with coordination, etc. These applications of refractometry are now only of historical interest. As the refraction of the oxide part of the silicate rests on N_c, refractometric methodologies are beneficial for analyzing vitreous or polycrystalline silicates to ascertaining the physical significance of anions and also to define N_c of cations.

REVIEW QUESTIONS:

1. Describe in detail the reflective index?
2. What are the structural applications of refractometry?
3. Describe the dipole moment mechanism?
4. What is polarization of dipole moments?

REFERENCES

1. Afromowitz, M. A., (1974). Refractive index of Ga1− xAlxAs. *Solid State Communications, 15*(1), 59–63.
2. Amos, A., & Burrows, B., (1973). Solvent-shift effects on electronic spectra and excited-state dipole moments and polarizabilities *Advances in Quantum Chemistry* (Vol. 7, pp. 289–313): Elsevier.
3. Banerjee, A., (2009). Enhanced refractometric optical sensing by using one-dimensional ternary photonic crystals. *Progress In Electromagnetics Research, 89*, 11–22.
4. Baonza, V., Taravillo, M., Caceres, M., & Nunez, J., (2006). Extension of the Szigeti equations: Average longitudinal-transverse frequencies and effective charges. *Physical Review B, 73*(21), 214117.
5. Barer, R., (1957). Refractometry and interferometry of living cells. *JOSA, 47*(6), 545–556.
6. Barth, C., Goncalves, M., Pires, A., Roeder, J., & Wolf, B., (2000). Asymmetric polysulfone and polyethersulfone membranes: Effects of thermodynamic conditions during formation on their performance. *Journal of Membrane Science, 169*(2), 287–299.
7. Batsanov, S. S., (1961). *Refractometry and Chemical Structure*: Consultants bureau.
8. Bin Mat Yunus, W. M., & bin Abdul Rahman, A., (1988). Refractive index of solutions at high concentrations. *Applied Optics, 27*(16), 3341–3343.
9. Bond, W., (1965). Measurement of the refractive indices of several crystals. *Journal of Applied Physics, 36*(5), 1674–1677.
10. Chen, Y., Liu, J., Yang, Z., Wilkinson, J. S., & Zhou, X., (2019). Optical biosensors based on refractometric sensing schemes: A review. *Biosensors and Bioelectronics, 144*, 111693.
11. Dorrah, A. H., Zamboni-Rached, M., & Mojahedi, M., (2018). Experimental demonstration of tunable refractometer based on orbital angular momentum of longitudinally structured light. *Light: Science & Applications, 7*(1), 40.
12. Edlén, B., (1966). The refractive index of air. *Metrologia, 2*(2), 71.
13. Everitt, P., Sooriyabandara, M., Guasoni, M., Wigley, P., Wei, C., McDonald, G., & Kuhn, C., (2017). Observation of a modulational instability in Bose-Einstein condensates. *Physical Review A, 96*(4), 041601.

14. Fern, R., & Onton, A., (1971). Refractive index of AlAs. *Journal of Applied Physics, 42*(9), 3499–3500.

15. Foteinopoulou, S., Economou, E. N., & Soukoulis, C., (2003). Refraction in media with a negative refractive index. *Physical Review Letters, 90*(10), 107402.

16. Grancini, G., Marras, S., Prato, M., Giannini, C., Quarti, C., De Angelis, F., & Manna, L., (2014). The impact of the crystallization processes on the structural and optical properties of hybrid perovskite films for photovoltaics. *The Journal of Physical Chemistry Letters, 5*(21), 3836–3842.

17. Hickey, A. L., & Rowley, C. N., (2014). Benchmarking quantum chemical methods for the calculation of molecular dipole moments and polarizabilities. *The Journal of Physical Chemistry A, 118*(20), 3678–3687.

18. Hush, M., Szigeti, S., Carvalho, A., & Hope, J., (2013). Controlling spontaneous-emission noise in measurement-based feedback cooling of a Bose–Einstein condensate. *New Journal of Physics, 15*(11), 113060.

19. Jasien, P. G., & Fitzgerald, G., (1990). Molecular dipole moments and polarizabilities from local density functional calculations: Application to DNA base pairs. *The Journal of Chemical Physics, 93*(4), 2554–2560.

20. Jensen, L., Åstrand, P. O., & Mikkelsen, K. V., (2001). An atomic capacitance–polarizability model for the calculation of molecular dipole moments and polarizabilities. *International Journal of Quantum Chemistry, 84*(5), 513–522.

21. Karlström, G., & Sadlej, A. J., (1982). Basis set superposition effects on properties of interacting systems. Dipole moments and polarizabilities. *Theoretica Chimica Acta, 61*, 1–9.

22. Kordes, E., (1940). Ionic radii and the periodic system, II. Calculation of ionic radii from atomic physical properties: *Zeitschr. Physik. Chem., v. B, 48*, 91–107.

23. Kumar, V., & Singh, J., (2010). *Model for Calculating the Refractive Index of Different Materials 1*(2), 13–22.

24. Li, F., Hou, X., Pan, S., & Wang, X., (2009). Growth, structure, and optical properties of a congruent melting oxyborate, $Bi_2ZnOB_2O_6$. *Chemistry of Materials, 21*(13), 2846–2850.

25. Liptay, W., (1974). Dipole moments and polarizabilities of molecules in excited electronic states *Excited States* (Vol. 1, pp. 129–229). Elsevier.
26. Liptay, W., Wortmann, R., Schaffrin, H., Burkhard, O., Reitinger, W., & Detzer, N., (1988). Excited state dipole moments and polarizabilities of centrosymmetric and dimeric molecules. I. Model study of a bichromophoric molecule. *Chemical Physics, 120*(3), 429–438.
27. Lyddane, R. H., Sachs, R. G., & Teller, E., (1941). On the polar vibrations of alkali halides. *Physical Review, 59*(8), 673.
28. Malham, I. B., & Turmine, M., (2008). Viscosities and refractive indices of binary mixtures of 1-butyl-3-methylimidazolium tetrafluoroborate and 1-butyl-2, 3-dimethylimidazolium tetrafluoroborate with water at 298 K. *The Journal of Chemical Thermodynamics, 40*(4), 718–723.
29. Maligne, D., & Bruneaux, G., (2011). *Time-Resolved Fuel Film Thickness Measurement for Direct Injection SI Engines Using Refractive Index Matching: SAE Technical Paper 1*(3), 1–25.
30. Marple, D., (1964). Refractive index of GaAs. *Journal of Applied Physics, 35*(4), 1241–1242.
31. Notomi, M., & Mitsugi, S., (2006). Wavelength conversion via dynamic refractive index tuning of a cavity. *Physical Review A, 73*(5), 051803.
32. Pan, Z., & Trusler, J. M., (2022). Refractive index effects in pendant drop tensiometry. *International Journal of Thermophysics, 43*(5), 65.
33. Phillips, J. C., & Van Vechten, J. A., (1969). Dielectric classification of crystal structures, ionization potentials, and band structures. *Physical Review Letters, 22*(14), 705.
34. Pitoňák, M., Holka, F., Neogrády, P., & Urban, M., (2006). Optimized virtual orbitals for correlated calculations: Towards large scale CCSD (T) calculations of molecular dipole moments and polarizabilities. *Journal of Molecular Structure: THEOCHEM, 768*(1–3), 79–89.
35. Ramakrishnan, M., Shrestha, A., & Soar, J., (2020). *Inclusion of Complementary Industry Knowledge in IT Service Management Curriculum – A Case Study 2*(5), 9–18.
36. Reis, J. C. R., Lampreia, I. M., Santos, Â. F., Moita, M. L. C., & Douhéret, G., (2010). Refractive index of liquid mixtures: Theory and experiment. *ChemPhysChem, 11*(17), 3722–3733.
37. Rong, Q., Qiao, X., Guo, T., Wang, R., Zhang, J., Hu, M., & Ma, Y., (2012). Temperature-calibrated fiber-optic refractometer based on a compact FBG-SMS structure. *Chinese Optics Letters, 10*(3), 030604.

38. Schroeder, K., Ecke, W., Mueller, R., Willsch, R., & Andreev, A., (2001). A fiber Bragg grating refractometer. *Measurement Science and Technology, 12*(7), 757.
39. Singh, S., (2002). Refractive index measurement and its applications. *Physica Scripta, 65*(2), 167.
40. Skafte-Pedersen, P., Nunes, P. S., Xiao, S., & Mortensen, N. A., (2009). Material limitations on the detection limit in refractometry. *Sensors, 9*(11), 8382–8390.
41. Smith, D. R., Pendry, J. B., & Wiltshire, M. C., (2004). Metamaterials and negative refractive index. *Science, 305*(5685), 788–792.
42. Sun, Y., Chen, X., Sun, L., Guo, X., & Lu, W., (2003). Nanoring structure and optical properties of Ga8As8. *Chemical Physics Letters, 381*(3–4), 397–403.
43. Surdo, S., Carpignano, F., Strambini, L., Merlo, S., & Barillaro, G., (2014). Capillarity-driven (self-powered) one-dimensional photonic crystals for refractometry and (bio)sensing applications. *RSC Advances, 4*(94), 51935–51941.
44. Szigeti, B., (1949). Polarisability and dielectric constant of ionic crystals. *Transactions of the Faraday Society, 45*, 155–166.
45. Targema, M., Obi-Egbedi, N. O., & Adeoye, M. D., (2013). Molecular structure and solvent effects on the dipole moments and polarizabilities of some aniline derivatives. *Computational and Theoretical Chemistry, 1012*, 47–53.
46. Tariq, M., Forte, P. A., Gomes, M. C., Lopes, J. C., & Rebelo, L. P. N., (2009). Densities and refractive indices of imidazolium-and phosphonium-based ionic liquids: Effect of temperature, alkyl chain length, and anion. *The Journal of Chemical Thermodynamics, 41*(6), 790–798.
47. Veselago, V., Braginsky, L., Shklover, V., & Hafner, C., (2006). Negative refractive index materials. *Journal of Computational and Theoretical Nanoscience, 3*(2), 189–218.
48. Vörös, J., (2004). The density and refractive index of adsorbing protein layers. *Biophysical Journal, 87*(1), 553–561.
49. Wang, P., Brambilla, G., Ding, M., Semenova, Y., Wu, Q., & Farrell, G., (2011). Investigation of single-mode–multimode–single-mode and single-mode–tapered-multimode–single-mode fiber structures and

their application for refractive index sensing. *JOSA B, 28*(5), 1180–1186.

50. Wanstall, C. T., Agrawal, A. K., & Bittle, J. A., (2020). Implications of real-gas behavior on refractive index calculations for optical diagnostics of fuel–air mixing at high pressures. *Combustion and Flame, 214*, 47–56.
51. Wasastjerna, J. A., (1939). On the structure of solid solutions. *Physical Review, 55*(10), 986.
52. Wemple, S. H., & DiDomenico Jr, M., (1971). Behavior of the electronic dielectric constant in covalent and ionic materials. *Physical Review B, 3*(4), 1338.
53. Wu, Q., Semenova, Y., Wang, P., & Farrell, G., (2011). High sensitivity SMS fiber structure-based refractometer – analysis and experiment. *Optics Express, 19*(9), 7937–7944.
54. Zhu, T., & Gao, S.-P., (2014). The stability, electronic structure, and optical property of TiO_2 polymorphs. *The Journal of Physical Chemistry C, 118*(21), 11385–11396.

CHAPTER 7

The Element Structure of Non-Metals

Contents

Unit Introduction 172
7.1. Hydrogen and the Halogens 172
7.2. Chalcogens 173
7.3. Elements of the Fifth Main Group 176
7.4. Elements of the Fifth and Sixth Main Groups Under Pressure 179
7.5. Carbon 181
7.6. Boron 185
7.7. Summary 187
Review Questions: 187
Multiple Choice Questions: 188
References 189

UNIT INTRODUCTION

The 8-N rule states that an atom X belonging to an element in the periodic table's N-th main group will take part in 8-N covalent bonds (N=4 to 7):

B [XX] = 8 N

Additionally, the maximization of connectivity is a general rule of elements in the third and higher periods, where multiple bonds in atoms are avoided and 8-N bonds are typically formed with 8-N different atoms. However, because carbon is a second-period element, fewer bonds mean graphite as one of the most stable compounds than diamond under normal circumstances. The concept of maximal connectivity becomes more significant at higher pressures, at which point, diamond gets more stable (Morf et al., 2007).

7.1. HYDROGEN AND THE HALOGENS

Even in their solid states, the elements including hydrogen gas, fluorine, bromine, chlorine, and iodine comprise X_2 molecules. Rotating H_2 molecules pack themselves into solid hydrogen in a hexagonal-closet pattern. In **α**-F_2, the F_2 molecules are arranged in hexagonal layers stacked similarly to cubic closest-packing. The molecules have orientations perpendicular to the layers. The modification **β**-F_2 is stable when it exceeds its melting point from 45.6 K to (53.5 K), in which the molecules rotate regarding about their centers of gravity (Fourmigué & Batail, 2004). Chlorine, bromine, and iodine molecules are organized differently in crystalline form. It is amazing how far apart the atoms of adjacent molecules are from one another. Compared to molecules in the next plane, some intermolecular connections in the b-c plane are shorter and use the van der Walls distance, which is seen in organic and inorganic compounds (Chudzinski, McClary, & Taylor, 2011). As a result, we see some halogen molecules interacting with one another in the b-c plane. From chlorine to iodine, this association gets stronger. The crystals plate-like habit and simple cleavage parallel to the layers manifest the weaker attractive forces that exist between planes. The heavy substances of the fifth and sixth are the primary group members exhibiting comparable association tendencies (Aakeröy, Schultheiss, Rajbanshi, Desper, & Moore, 2009).

The arrangement of the halogen atoms makes sense due to the cubic closest packing created due to covalent interactions between molecules. Pressure reduce the deformations and lessens the contact lengths between the atoms. Iodine approximates continuously with pressure; but at P = 23 GPa, an abrupt stage where shift can occur leading to the incommensurately altered crystallized structure of I-V. In this instance, the F mmm ($00q_3$) 0s0 with q_3 = 0.257 at 24.6 GPa gives the four-dimensional superspace group. Despite the atom dispersion and sine wave-like motion along c, the structure may be estimated in three dimensions (as an average of the pattern) in the orthorhombic arrangement of spaces F mmm (Awwadi, Taher, Haddad, & Turnbull, 2014).

7.2. CHALCOGENS

In its solid state, oxygen is made up of O_2 molecules. They have a packing based on σ-F_2 between 24 K and 43.6 K. The packing is correspondingly noticeable at room temperature and pressure (5.5 GPa). The molecules lean on the hexagonal layer slightly below 24 K. The particles turn in the crystal according to β-F_2 (43.6 K) and reach to the melting point temperature (54.8 K). At about 100 GPa, oxygen turns metallic under pressure, but it remains a molecule (Vogel, Wonner, & Huber, 2019).

Remember

Alkali metals and graphite combine to form intercalation compounds.

Sulfur exhibits the most range of structural changes of any element. Crystal configurations are recognized for the subsequent forms: S_6, S_7 (four variations), also including S_6. S_{10}, S_8, S_{10}, S_{11}, S_{12}, S_{13}, S_{14}, S_{15}, and S_{18} (two forms), S_{20}. Among them can be generated using specific chemical synthesis, or they can be extracted from concepts that originated by extracting sulfur melt that had been quenched. One among the polymeric forms created by quenched sulfur melts has been identified by its molecular composition (Chivers & Laitinen, 2015). The rings or chain that makes up the aforementioned sulfur forms each include two sulfur atoms that adhere to 8-N rule. The disulfide bond (S-S bond) lengths typically are about 206 pm while exhibits some disseminate (Ho, Wang, Meloni, & Vargas-Baca, 2020). The dihedral angles vary from 74 to 100° and the S-S-S bond angles from 101 to 110°. A series of five atoms may therefore choose one of two arrangements:

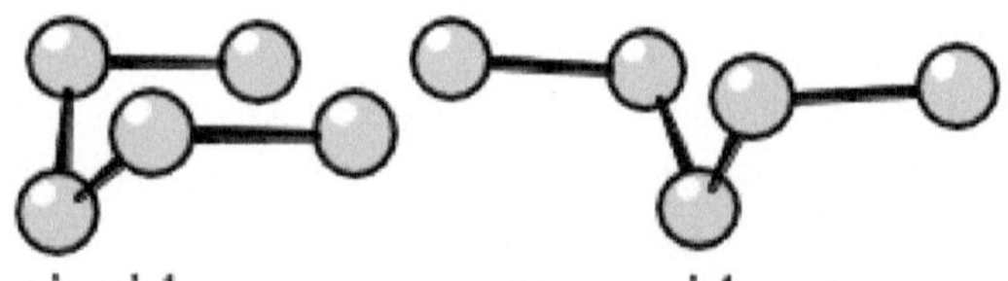

The smaller rings such as S_6, S_7, and S_8 only have cisoid arrangement because the dihedral angles had to adjust (74.5° for S and 98° for S). S has a chair conformation, and S_8 has a crown-form conformation. If one S atom were removed from S_8, S_7 could theoretically be created. In order for larger rings to be strain-free, cisoid and transoid groups must be present. Transoid and cisoid groups alternate in S_{12}. Only transoid groups can form helical chains, and the number of turns essential to effect an atom directly above one arranged a line perpendicular to the helical structures depends on the dihedral angle. It requires 10 atoms over three turns in one type of polymer made of sulfur (Zannoni, Borsetti, Harrison, & Turner, 2007).

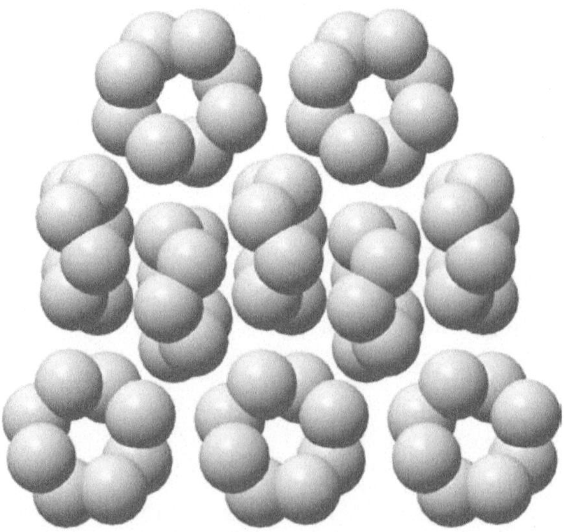

Figure 7.1. Section of the structure of α-sulfur.

Source: Wikipedia, Creative Commons License.

S_8 rings are arranged to form columns in orthorhombic **α**-sulfur, which makes the change stable under normal circumstances. In order to give the column, the appearance of a crank shaft, the consecutive rings are staggered rather than stacked one exactly on top of the other (as on a roll of coins). Columns in two opposite directions enable densely packing of the molecules in this configuration. The columns facing one direction are positioned in the 'crank shafts' at right angles. The rings are precisely stacked one on top of the other in S_6 and S_{12}, and the rolls are arranged parallel. A universally true norm can be seen in those structural molecules tend to pack as firmly as seen in the solid state (Mutra, Kudale, Li, Tsai, & Wang, 2020) (Figure 7.1).

Three selenium modifications with Se_8 rings are recognized. The rings are stacked differently than in the S_8 modifications because they look like coin rolls instead of being level. Additionally, a modification made of Se_6 molecules is identical to S_6. Three Se atoms are found in each turn of the helical chains that make up **α**-selenium, also known as Se-I, the form of selenium that is thermodynamically stable (Bleiholder, Werz, Köppel, & Gleiter, 2006). The crystal has chains arranged in parallel. Four elements situated 344 pm apart and belonging to three separate chains encircle each selenium atom. Approximate distance of 237 pm, the two adjacent atomic particles in the sequence and an octahedral shape contains 2 + 4 coordination are created. The Se–Se distance among the series cannot be predicted with adequate precision in the van der Waals separation (Benz, López□Andarias, Mareda, Sakai, & Matile, 2017) (Figure 7.2).

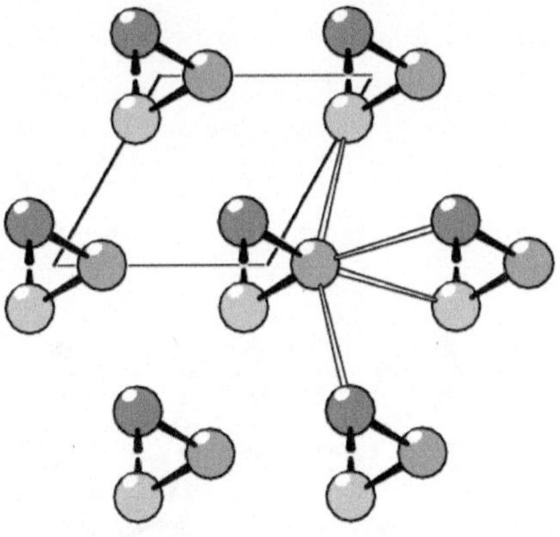

Figure 7.2. Structure of α-selenium.

Source: Wikipedia, Creative Commons License.

Tellurium is similar to α-selenium when it crystallizes. As one might anticipate, the chain's Te-Te interactions (283 pm) are larger than those of the selenium element but its lengths in the atoms of the subsequent series (Te…Te 349 pm) are nearly identical. In regard to the van der Waals distance, reduction in the atomic divergence from their typical octahedral arrangement and the shortening are more evident. Applying pressure will equalize all six distances (Kolb, Oliver, & Werz, 2020). For polonium, there are two known modifications. The cubic-primitive structure of α-polonium, which has a precise octahedral coordination for each atom, makes it stable at room temperature. At high pressures, this structure also appears for phosphorus and antimony, though it is somewhat unusual. α-Po is changed to β-Po at 54°C. Rhombohedral lattice is produced due to the phase transition, which involves compression along the cubic unit cell body diagonals. Bond angles total 98.2° (Pascoe, Ling, & Cockroft, 2017).

7.3. ELEMENTS OF THE FIFTH MAIN GROUP

In solid nitrogen-containing compounds, there exist five recognized variations in terms of the arrangement of Nitrogen molecules. With a transition temperature of 35.6 K, only two of them may exist at the pressure of the atmosphere, while the others require extremely very high pressure. At a pressure of approximately 100 GPa, a change in phase with audible hysteresis will occur, leading to a non-molecular modification (Miyoshi, Sakai, Tanaka, & Masamura, 1998). This probably belongs to a type that contains arsenic. Electrical conductivity becomes apparent around 140 GPa. At higher temperatures, phosphorus vapor includes P_2 atoms, which are tetrahedral P_4 molecules. White phosphorus is produced by condensing the vapor; it additionally contains P4 molecules. P_4 molecules make up liquid phosphorus in its natural state, but at 1 GPa and 100°C, polymeric liquid state of phosphorus is formed that is incompatible to liquid P_4 (Rosenbaum, Rubin, & Sandberg, 1940) (Figure 7.3).

Figure 7.3. The structure of black phosphorus.

Source: Wikipedia, Creative Commons License.

White phosphorus can be converted to red phosphorus by heating it to temperatures over 180°C or exposure to light. The conditions of preparation impact the color, melting point, vapor pressure, and particularly density of the substance. It is typically amorphous or microcrystalline, and crystal growth is rather difficult (Kutzelnigg, 1984). At temperatures around 550°C, fibrous red phosphorus and platelets count of Hittorf's (violet) phosphorus gradually takes a crystallize form. A solution in liquid lead was slowly cooled (from 630 to 520°C) to form only single crystals of Hittorf phosphorus. Both modifications consist of pentagonal-shaped polymeric tubes made of cages identical to those in As_4S_4 and As_4S_5 and connected by P_2 dumbbells. The tubes are associated in parallel pairing in stringy phosphorus. They are crosswise connected to grids in Hittorf phosphorus; pairs of grids overlap but not bonded to one another. Every P atom connects to three additional atoms according to the 8-N rule (Höllwarth et al., 1993).

Despite having a complex structure, these modification linking principle frequently appears inside the structural composition of phosphorus compounds. Between the polyphosphides and polyphosphanes, there are known building blocks that correspond to pieces of tubes. $P_{15}Se$ and $P_{19}Se$ also contain tubes similar to these. At 550°C, (CuI) substance can be made from the various elements. It has a polymeric phosphorus arrangement that can be secluded after the copper iodide has been extracted using a potassium cyanide aqueous solution. The molecules are made up of P_8 cages connected by P_4 squares. Similar steps can be taken to produce a different variety from the $(CuI)_3P_{12}$; its sequences are made up of P_{10} cages and P_2 dumbbells (Atanasov & Reinen, 2005).

Only under very precise conditions does black phosphorus occur (greater pressure, crystallization from liquid state elements like bismuth or extended boiling in the existence of mercury Hg), although in most cases, this modification is the most thermodynamically stable one. It consists of layers that are placed in the form of a chair by six-membered rings. Ring pairs are linked, much similar to the rings in cis-decalin (Clavería-Cádiz, Arratia-Pérez, & MacLeod-Carey, 2016). A layer may additionally be viewed by means of an assembly of interconnected zigzag lines that switch between double separate planes. Each P atom in a layer is joined to three more P ions by bonds that are separated by 222 and 224 pm. The van der Waals separation of the atomic distances is 2*359 pm and 1*380 pm between the layers. The polyphosphides also contain some of the black phosphorus' structural characteristics (Bergner, Dolg, Küchle, Stoll, & Preuß, 1993) (Figure 7.4).

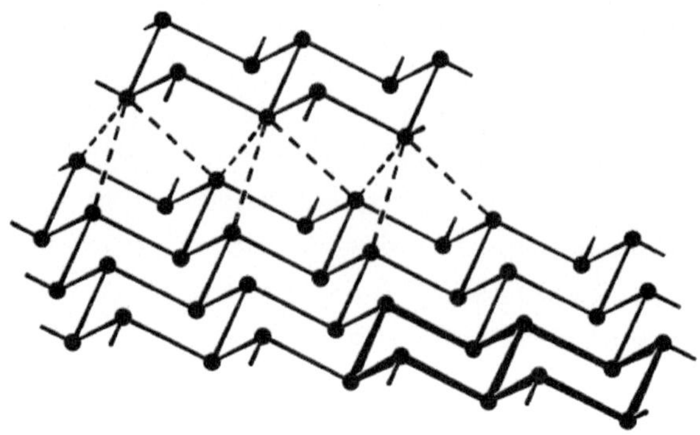

Figure 7.4. Section of a layer in grey arsenic.

Source: Wikipedia, Creative Commons License.

Arsenic has been altered using white and black phosphorus crystals. But grey (metallic, rhombohedral) is the only stable form of arsenic. Similar to trans-decalin, it is composed of layers of chair-shaped six-membered rings which are joined to one another. The layer's atoms continuously reside in a higher and lower plane. Every ring has an As atom in a neighboring layer over and under its center because of the layers' staggered stacking. Each As atom has a distorted octahedral 3 + 3 coordination, which means that in addition to the three atoms it is connected to inside the layer, it is also in contact with a further three atoms. As-As bonds in layers have a length of approximately 252 pm, while the distance between particles in adjacent layers reaches 312 pm, that close together rather than to the van der Waals distance having 370 pm (Wadt & Hay, 1985).

The structures of bismuth and antimony are similar to those of grey arsenic. In other words, as atomic weight increases, coordination in polyhedral structures diverge from regular octahedral compounds less because there is less fluctuation in the spacing of atoms adjacent inside and between layers. This effect is amplified while under pressure. When the atomic structures for example, As, P, Sb, and Bi are denoted as layer structures and structural appearance of Se and Te denoted as chain arrangements, the occurrence of bonding interactions between the layers arrangement and chains is unnoticed (Grund, Schuster, & Mueller, 1984). These interfaces grow more distended for the denser atoms. In the case of Sb and Bi, wherein the interlayer distances between neighboring atoms are just 15% larger than the inter layer distances, the definite displacement from the traditional polonium arrangement is somewhat minor. Conductivity in metallic elements is also present among Sb, As, and Bi. The bonding interactions can be understood using band theory. A Peierls deformation begins with the α-Po structure and strengthens three bonds per atoms. Tellurium and selenium have stronger double bonds than other elements (Dodds, Jagoda, Reglinski, & Spicer, 2004) (Table 7.1).

Table 7.1. Distances between Adjacent Atoms and Bond Angles in Structures of the a-As, a-Se, a-Po and b-Po Type

Condition	Structure Type	d1	d2	d2d1	An-gle
P(10 GPa)	a-As	213	327	1.54	105
P(12 GPa)	a-Po	238	238	1.00	90
As	a-As	252	312	1.24	96.6
As(25GPa)	a-Po	255	255	1.00	90
Se	a-Se	237	344	1.45	103.1
Sb	a-As	291	336	1.15	95.6
Te	a-Se	283	349	1.23	103.2
Te-IV(11.5GPa)	b-Po	295	295	1.00	102.7
Bi	a-As	307	353	1.15	95.5
Po	a-Po	337	337	1.00	90
Po	b-Po	337	337	1.00	98.2

*d_1 is bond distance; d_2 is shortest interatomic distance between layers or chains; distances in pm, angles in degrees.

7.4. ELEMENTS OF THE FIFTH AND SIXTH MAIN GROUPS UNDER PRESSURE

The extremely high levels of the X-rays have made it possible to determine crystal structure from very small samples. A small sample sandwiched between two diamond anvils can withstand extremely high pressures. This has greatly expanded our understanding of how matter behaves under high pressure. The fifth and sixth main group elements display somewhat peculiar structures when stressed (Mann, Meek, & Allen, 2000).

A tellurium atom's natural coordination is 2 + 4, which is in the case of elemental tellurium. It has long been understood that interatomic distances gradually decrease under pressure until, at 297 pm, every tellurium atom has six evenly spaced neighbors; this structure, now known as Te-IV, is equivalent to -polonium. Te-II and Te-III, two more, unexpected alterations, first arise at 4 GPa and 7 GPa before being accomplished. Te-II is made up of parallel, straight series that are simultaneously lifted in such a method in which Te atom, in addition to its two chain mates (310 pm), has two neighbors that are closer to it (286–299 pm) and four farther away (331–364 pm) (Wiberg, 1997). Each tellurium atom in Te-III has six atoms that are close by, located between 297 and 316 pm, and six more that are far away, located between 368 and 392 pm. Te-III has a structural change that is not proportional. Tellurium finally transitions into a typical metallic structure, the body-centered cubic arrangement (Te-V), at a pressure of 27 GPa.At least, five highest pressure of sulfur element alterations exist, in which one of them (>80 GPa) has the primary-polonium assembly (Loose et al., 1999) (Figure 7.5).

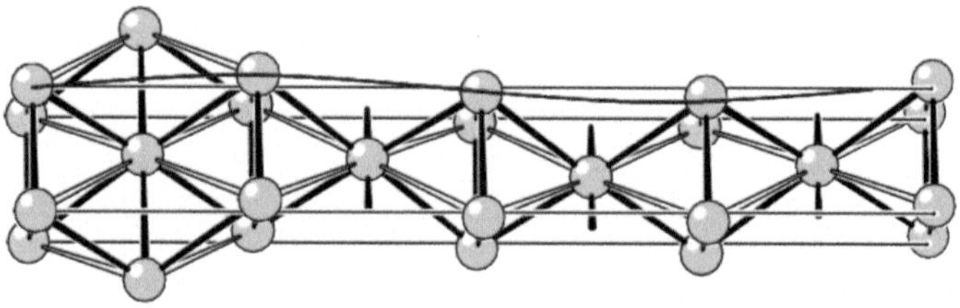

Figure 7.5. Four body-centered unit cells of the incommensurately modulated structure of tellurium-III.

Source: Wikipedia, Creative Commons License.

Phosphorus black first undergoes a variation under pressure that is analogous to gray arsenic. This changes into the **β**-polonium structure at a higher pressure. The next structure is a primitive hexagonal structure, which is extremely rare otherwise but has been seen for silicon under pressure. Phosphorus is body-centered cubic above 262 GPa; below 22 K, this modification turns into superconductivity. At 25 GPa, arsenic also takes on the -polonium structure, and at the highest pressures, it transforms into body-centered cubic. Between these two modifications, there is an intermediate structure that is rather unusual Bi-III (Weetman & Inoue, 2018).

Between 2.8 and 8 GPa, bismuth exhibits this Bi-III structure, as does antimony at 10 and 28 GPa. Antimony and bismuth assume the cubic body-centered organization of sphere of influence that is characteristic for various metals at even higher pressures. The Bi-III's composite crystal structure is peculiar and discordant. It can be designated by double partly intergrown structures that cannot be measured together. The partial structure is made up of square antiprisms that are joined by tetrahedral elements and have one face each along c. Partial framework 2 produces atoms which connect together c in a series of linear chains inside center of the square antiprisms. In order to compensate for various distances among various atoms within the chains and the remaining antiprisms, both partial structures are likewise incommensurately modulated. The atomic particles in the chains are somewhat off-center along c, whereas the atoms of the antiprism are perpendicular to c (Igel-Mann, Stoll, & Preuss, 1988).

In conclusion, it can be said that the pressure required to produce a typical metallic structure decreases with increasing atomic number. Unusual structures that cannot be incorporated into conventional chemical models appear halfway between non-metallic and metallic structures (Wadt & Hay, 1985) (Figure 7.6).

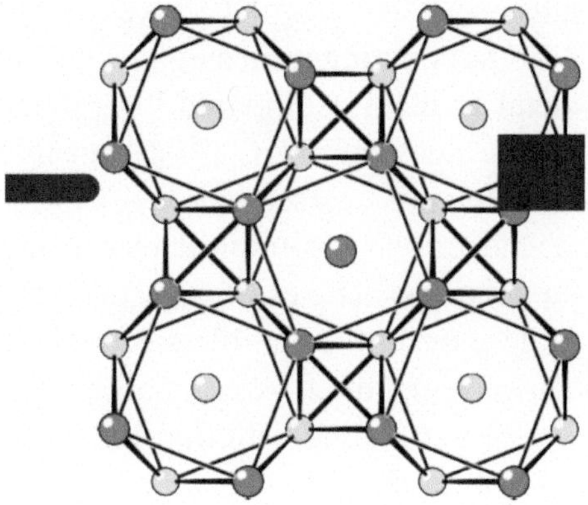

Figure 7.6. The incommensurate structure of Bismuth-III.

Source: Wikipedia, Creative Commons License.

7.5. CARBON

The carbon modification known as graphite is stable in typical circumstances. It is made up of planar layers. Each C atom in the layer is covalently bonded to three other C atoms to the layer's delocalized bond system. Because of this, the band is only half full, and the material is in a metallic state that conducts electricity in two dimensions. Weaker van der Waals forces are most common attractive forces within the strata. Layers are separated by 335 pm, and the linkages that connect them are 142 pm long (Buckley, 1988). Since the layers are not parallel, the high electrical conductivity exists only perpendicular to them. The atomic particles contain at least at one layer are placed halfway directly above those in the layer beneath it, also those of the other layer are placed precisely above the rings' centers. A, B, and C are the three potential layer placements. Although sections with the sequence ABC typically separate regions with the prevailing sequence ABAB… in normal (hexagonal) graphite, this more or less empirical layer sequence is also common. One-dimensional disorder is what this is, where the periodic order is absent in the direction of stacking, but the atoms are ordered within the layers (Lehmann, 2007).

Did You Know?

No other element displays as many structural variations as sulfur.

Alkali metals and graphite combine to form intercalation compounds. They are made of substances like KC_8, KC_{24}, KC_{36}, or KC_{48}, or LiC_6, LiC_{12}, or LiC_{18}. They come in hues ranging from a golden luster to black, depending on the amount of metal in them. Compared to graphite, they are superior electric conductors. In the elements of KC_8 (first stage intercalation), between each pair of graphite layers, in KC_{24} (second stage), etc., the alkali metal ions are intercalated. The valence band of graphite receives the valence electrons that belong to the metal atoms. The electrodes of lithium-ion batteries use the potential for the more or less reversible electrochemical insertion in Li ions to graphite in various quantities. Among all, metal chlorides MCl_n (M = almost all metals; n = 2 to 6) along with fluorides also in the bromides are another type of intercalation compound. The structures of the intercalated halide sections essentially match those of the pure compounds; for instance, the structure of intercalated $FeCl_3$ layers matches with $FeCl_3$, as depicted on the obverse cover (Sheehan, Buesking, & Sullivan, 1994).

In theory, carbon in all of its forms, including fossil fuels, coke, burning charcoal, soot, etc., is comparable to graphite even though it has a relatively small amount of ordering. The ends of the microcrystalline or amorphous graphite layer fragments are joined by OH groups and maybe other atom groups (Haddon, 2002). Because of their numerous pores, many carbon forms have vast interiors which allow them to act as catalysts and absorb significant amounts of other compounds. In this aspect, crystalline graphite is less active. When polyacrylonitrile fibers are pyrolyzed, carbon fibers are created from layers of graphite that run opposite to the direction of the fiber (Edwards, Marsh, & Menendez, 2013).

Carbon modifications called fullerenes are made of cage-like molecules. It is possible to create them by establishing an electric curvature among two graphite conductors in a regular structured helium environment, reducing the carbon after evaporation, and then changed from re-crystallized state to a magenta-colored benzene solution. Buckminsterfullerene, also known as fullerene C_{60}, is the main product. The C_{60} molecule is shaped like a soccer ball and has 20 hexagons resembling benzene and 12 pentagons. C_{70}, which has a shape resembling a peanut and 12 pentagons and 25 hexagons, comes in second place in terms of yield from this preparation. Although it is possible to make cages of other sizes (starting at C_{32}), they are less stable and can contain an even number of C atoms. A fullerene molecule constantly contains 12 pentagons, regardless of size (Windhorst & Blount, 1997) (Figure 7.7).

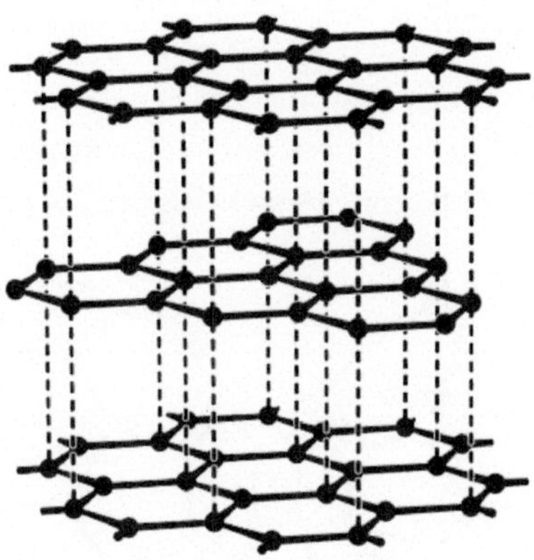

Figure 7.7. The structure of graphite (stereo-image).

Source: Wikipedia, Creative Commons License.

Crystalline C_{60} has a cubic face-centered arrangement, also known as a cube closest-packing of spheres. The molecules spin inside the crystal because they are almost spherical. The crystals are delicate, like graphite. Similar to the intercalation complexes of graphite, which potassium atoms can be contained. They fill the spaces within the C_{60} balls. When all cavities (including the tetrahedral and octahedral voids if the C_{60} balls are believed to be closely packed spheres) are filled, the composition is K_3C_{60}. This chemical, which exhibits metallic properties, becomes a superconductor when cooled below 18 K. Extra potassium ions can be interspersed because of the body-centered cubic pack of the C_{60} atoms in K_6C_{60} (Buckley & Edie, 1993).

7.5.1. Carbon Nanotubes

With the use of a laser or electric arc, graphite can be transformed into carbon nanotubes. They resemble tangled tubes. By initiating the burning of gaseous hydrocarbons at temperatures between 700 and 1,100°C, it is possible to create arranged collections made up of carbon nanotube on suitable carrier materials. Examples include the formation of single-walled nanotubes with an average diameter of 1.4 nm from the pyrolysis of acetylene in a ferrocene solution at 1,100°C and the formation of multi-walled nanotubes from the decomposition of benzene in a solution of $Fe(CO)_5$. Layers of bending graphite that have a length of 0.1 mm are known as nanotubes. The ends of the tubes are often sealed by a half fullerene sphere, which can be eliminated with ultrasound while suspended in an acidic liquid. The six-membered rings' orientations in relation to the tube axis can vary, affecting the electric conductivity. Tubes are arranged concentrically around one another to form multilayered nanotubes (Ebbesen, 1994) (Figures 7.8 and 7.9).

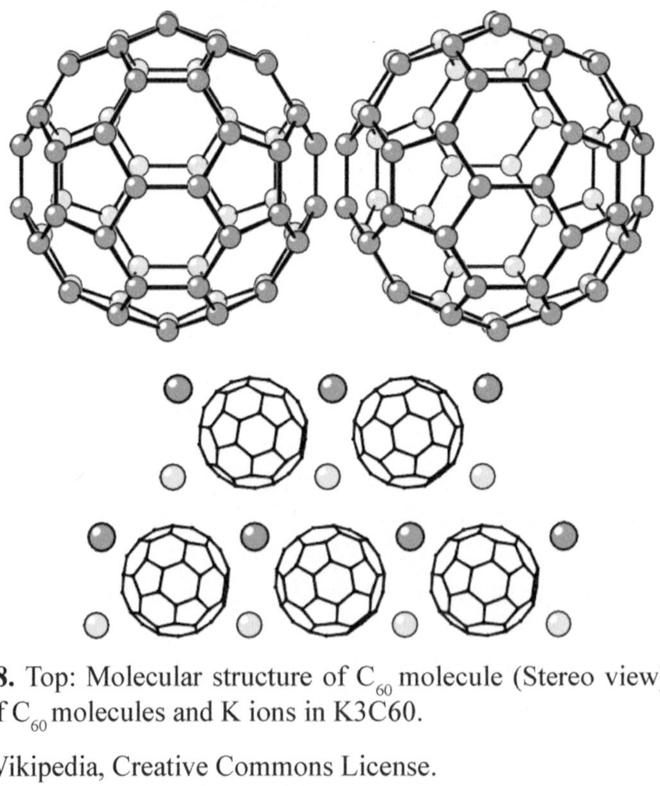

Figure 7.8. Top: Molecular structure of C_{60} molecule (Stereo view). Bottom: Packing of C_{60} molecules and K ions in K3C60.

Source: Wikipedia, Creative Commons License.

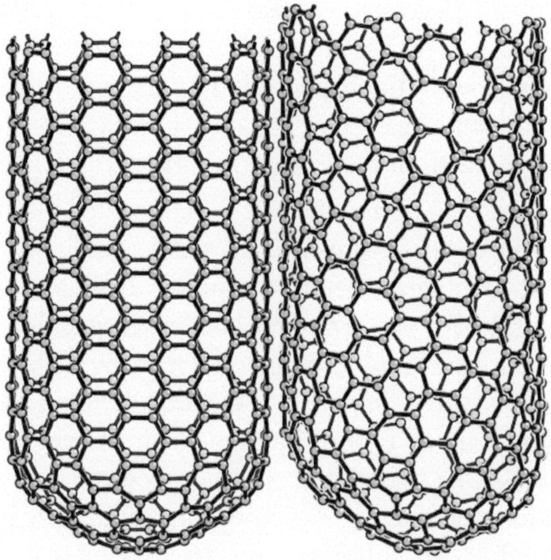

Figure 7.9. Structures of two kinds of mono-layered carbon nanotubes. The left one shows metallic conduction, the right one is semi-conducting.

Source: Wikipedia, Creative Commons License.

7.6. BORON

Both the chemical behavior and the structure of boron are peculiar. There are about 16 boron changes that have been reported, yet most have not been thoroughly considered. Many examples that were thoughts that are made entirely of boron may actually have been boron-rich borides (a lot of which are understood, such as YB_{66}). Rhombohedral -B_{12} is a recognized structure in which the subscript letter denotes the numeral of atoms in a single unit cell. Three additional forms, tetragonal B_{50}, rhombohedral B_{105}, and rhombohedral B_{320}, have known crystal structures, but it is likely that boron-rich borides were the focus of the research. $B_{48}X_2$ should be used to formulate **α**-B_{50}. It comprises B_{12} icosahedra connected by X atoms with tetrahedrally coordinated orbitals (Keren, 1996).

These atoms are most likely C or N atoms since X-ray diffraction makes it difficult to distinguish between B, C, and N. The notable structural element in all the different boron changes that have been characterized is the B_{12} icosahedron, which is also present in the anionic closo-borane $B_{12}H^2{}_{12}$. The 12 elements of an icosahedron are joined by multicenter bonds (Lovatt & Dugger, 1984). The MO theory predicts that there ought to be 13 bonding orbitals, each with 26 electrons, yet there are only 10 electrons with valence left. In the $B_{12}H^{2}_{12}$ ion, which is employed for almost 12 covalent B-H bonds that are directly concentrating and radially outward from the icosahedron shapes, there are a total of 24 electrons, or 12 electron pairs (12 from the Hydrogen atoms and 2 from the ionizing charges) (Bingham, 1983).

The B_{12} icosahedra in boron element are connected by such radial bonds, but not all of them can be the standard two-center, two-electron connections as there are only 10 valence electrons that are available for 12. In the B_{12} structure, the icosahedra are arranged in a cubic packing of spheres. Six additional icosahedra surround each icosahedron in one layer of icosahedra, which are joined by three-center, two-electron bonds (Parks & Edwards, 2005). These bonds involve boron atoms, each of which contributes two electrons on average, or two, three, six, four electrons in each icosahedron molecule. Additionally, each icosahedron is surrounded by six additional icosahedra from the two following levels, to which it is connected by conventional B-B bonds, necessitating each icosahedron to have six electrons. This sums up exactly to the 10 electrons mentioned above that make up the inter-icosahedron bonds (Loomis & Durst, 1992) (Figure 7.10).

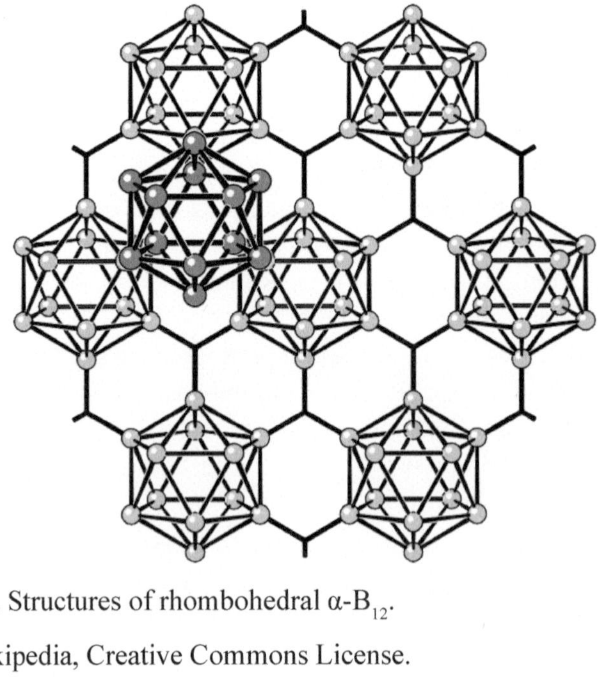

Figure 7.10. Structures of rhombohedral α-B_{12}.

Source: Wikipedia, Creative Commons License.

7.7. SUMMARY

In accordance with the 8-N rule, an atom X from the elements in the periodic table that has the N-th major group will participate in 8N covalent bonding. The elements including hydrogen, bromine chlorine, fluorine, and iodine make up molecules X_2 even in their solid states. O_2 molecules make up oxygen when it is solid. They pack as in α-F_2 and range in temperature from 24 K to 43.6 K. Sulfur contains more structural changes than any other element. Three selenium modifications with Se_8 rings are recognized. For polonium, there are two known modifications. There are five known modifications for solid nitrogen that vary in how the N_2 molecules are packed. Black phosphorus forms only under very specific circumstances (as due to crystallization from liquid bismuth, high pressure, or heating in the presence of Hg for long period of time). Antimony and bismuth have structures that are similar to gray arsenic. The extremely high levels of the X-rays have made it possible to determine crystal structure from very small samples. Black phosphorus first undergoes a modification under pressure that is analogous to gray arsenic. The carbon modification known as graphite is stable in typical circumstances. Alkali metals and graphite combine to form intercalation compounds. Carbon modifications called fullerenes are made of cage-like molecules. Graphite can be converted into carbon nanotubes using a laser or an electric arc. Both the chemical behavior and the structure of boron are peculiar. Many samples that were thought to be made entirely of boron may actually have been boron-rich borides.

Activity 6.1

Compare the structures of different non-metals using diagrams.

REVIEW QUESTIONS:

1. Write a detailed note on the structure of hydrogen and halogens.
2. Explain the structure of elements of 5th group in detail.
3. Explain in detail the structure of chalcogens.
4. What is the structure of carbon?
5. What is the structure of boron?

MULTIPLE CHOICE QUESTIONS:

1. Which of the following elements shows most of different structures?
   a. Carbon
   b. Hydrogen
   c. Oxygen
   d. Sulfur
2. How many modifications are formed by selenium?
   a. 2
   b. 3
   c. 4
   d. 5
3. How many modifications are formed by solid nitrogen?
   a. 2
   b. 3
   c. 4
   d. 5
4. Graphite is the modification of which of the following elements:
   a. Carbon
   b. Hydrogen
   c. Nitrogen
   d. Sulfur
5. How many modifications are formed by boron:
   a. 15
   b. 16
   c. 17
   d. 18

Answers to Multiple Choice Questions:

1. (d); 2. (b); 3. (d); 4. (a); 5. (b)

REFERENCES

1. Aakeröy, C. B., Schultheiss, N. C., Rajbanshi, A., Desper, J., & Moore, C., (2009). Supramolecular synthesis based on a combination of hydrogen and halogen bonds. Crystal Growth and Design, 9(1), 432–441.
2. Atanasov, M., & Reinen, D., (2005). The stability and eventual lone pair distortion of the hexahalide complexes and molecules of the fifth to eighth main-group elements with one lone pair, as isolated entities or in oligomeric clusters: A vibronic coupling and DFT study. Inorganic Chemistry, 44(14), 5092–5103.
3. Awwadi, F. F., Taher, D., Haddad, S. F., & Turnbull, M. M., (2014). Competition between hydrogen and halogen bonding interactions: Theoretical and crystallographic studies. Crystal Growth & Design, 14(4), 1961–1971.
4. Benz, S., López-Andarias, J., Mareda, J., Sakai, N., & Matile, S., (2017). Catalysis with chalcogen bonds. Angewandte Chemie International Edition, 56(3), 812–815.
5. Bergner, A., Dolg, M., Küchle, W., Stoll, H., & Preuß, H., (1993). Ab initio energy-adjusted pseudopotentials for elements of groups 13–17. Molecular Physics, 80(6), 1431–1441.
6. Bingham, F. T., (1983). Boron. Methods of Soil Analysis: Part 2 Chemical and Microbiological Properties, 9, 431–447.
7. Bleiholder, C., Werz, D. B., Köppel, H., & Gleiter, R., (2006). Theoretical investigations on chalcogen–chalcogen interactions: what makes these nonbonded interactions bonding? Journal of the American Chemical Society, 128(8), 2666–2674.
8. Buckley, J. D., & Edie, D. D., (1993). Carbon-Carbon Materials and Composites: William Andrew, 1(5), 10–26.
9. Buckley, J. D., (1988). Carbon-carbon-an overview. American Ceramic Society Bulletin, 1(67).
10. Chivers, T., & Laitinen, R. S., (2015). Tellurium: a maverick among the chalcogens. Chemical Society Reviews, 44(7), 1725–1739.
11. Chudzinski, M. G., McClary, C. A., & Taylor, M. S., (2011). Anion receptors composed of hydrogen-and halogen-bond donor groups: Modulating selectivity with combinations of distinct noncovalent interactions. Journal of the American Chemical Society, 133(27), 10559–10567.

12. Clavería-Cádiz, F., Arratia-Pérez, R., & MacLeod-Carey, D., (2016). Density functional study on Keggin heteropolyanions containing fifth period main group heteroatoms. Polyhedron, 117, 478–486.
13. Dodds, C. A., Jagoda, M., Reglinski, J., & Spicer, M. D., (2004). Soft scorpionate complexes of the fourth and fifth period elements arsenic and tin. Polyhedron, 23(2–3), 445–450.
14. Ebbesen, T. W., (1994). Carbon nanotubes. Annual Review of Materials Science, 24(1), 235–264.
15. Edwards, I. A., Marsh, H., & Menendez, R., (2013). Introduction to Carbon Science. Butterworth-Heinemann.
16. Fourmigué, M., & Batail, P., (2004). Activation of hydrogen-and halogen-bonding interactions in tetrathiafulvalene-based crystalline molecular conductors. Chemical Reviews, 104(11), 5379–5418.
17. Grund, I., Schuster, H., & Mueller, P., (1984). Ternary compounds of lithium with yttrium, lanthanum, or neodymium and elements of the fifth main group in a'filled'CaAl/sub 2/Si/sub 2/-type structure. Z. Anorg. Allg. Chem. (German Democratic Republic), 515.
18. Haddon, R. C., (2002). Carbon Nanotubes (Vol. 35, pp. 997–997): ACS Publications.
19. Ho, P. C., Wang, J. Z., Meloni, F., & Vargas-Baca, I., (2020). Chalcogen bonding in materials chemistry. Coordination Chemistry Reviews, 422, 213464.
20. Höllwarth, A., Böhme, M., Dapprich, S., Ehlers, A., Gobbi, A., Jonas, V., & Frenking, G., (1993). A set of d-polarization functions for pseudo-potential basis sets of the main group elements Al–Bi and f-type polarization functions for Zn, Cd, Hg. Chemical Physics Letters, 208(3–4), 237–240.
21. Igel-Mann, G., Stoll, H., & Preuss, H., (1988). Pseudopotentials for main group elements (IIIa through VIIa). Molecular Physics, 65(6), 1321–1328.
22. Keren, R., (1996). Boron. Methods of Soil Analysis: Part 3 Chemical Methods, 5, 603–626.
23. Kolb, S., Oliver, G. A., & Werz, D. B., (2020). Chemistry evolves, terms evolve, but phenomena do not evolve: From chalcogen–chalcogen interactions to chalcogen bonding. Angewandte Chemie International Edition, 59(50), 22306–22310.
24. Kutzelnigg, W., (1984). Chemical bonding in higher main group

elements. Angewandte Chemie International Edition in English, 23(4), 272–295.

25. Lehmann, J., (2007). A handful of carbon. Nature, 447(7141), 143–144.
26. Loomis, W. D., & Durst, R., (1992). Chemistry and biology of boron. BioFactors (Oxford, England), 3(4), 229–239.
27. Loose, I., Droste, E., Bösing, M., Pohlmann, H., Dickman, M. H., Rosu, C., & Krebs, B., (1999). Heteropolymetalate clusters of the subvalent main group elements BiIII and SbIII. Inorganic Chemistry, 38(11), 2688–2694.
28. Lovatt, C. J., & Dugger, W., (1984). Boron. Biochemistry of the Essential Ultratrace Elements, 389–421.
29. Mann, J. B., Meek, T. L., & Allen, L. C., (2000). Configuration energies of the main group elements. Journal of the American Chemical Society, 122(12), 2780–2783.
30. Miyoshi, E., Sakai, Y., Tanaka, K., & Masamura, M., (1998). Relativistic dsp-model core potentials for main group elements in the fourth, fifth and sixth row and their applications. Journal of Molecular Structure: TheoChem, 451(1–2), 73–79.
31. Morf, L. S., Tremp, J., Gloor, R., Schuppisser, F., Stengele, M., & Taverna, R., (2007). Metals, non-metals and PCB in electrical and electronic waste – Actual levels in Switzerland. Waste Management, 27(10), 1306–1316.
32. Mutra, M. R., Kudale, V. S., Li, J., Tsai, W.-H., & Wang, J.-J., (2020). Alkene versus alkyne reactivity in unactivated 1, 6-enynes: regio-and chemoselective radical cyclization with chalcogens under metal-and oxidant-free conditions. Green Chemistry, 22(7), 2288–2300.
33. Parks, J. L., & Edwards, M., (2005). Boron in the environment. Critical Reviews in Environmental Science and Technology, 35(2), 81–114.
34. Pascoe, D. J., Ling, K. B., & Cockroft, S. L., (2017). The origin of chalcogen-bonding interactions. Journal of the American Chemical Society, 139(42), 15160–15167.
35. Rosenbaum, E., Rubin, D. J., & Sandberg, C. R., (1940). The Raman spectra of phosphorus methyl and arsenic methyl and the force constants of the methyl compounds of the fifth group elements. The Journal of chemical physics, 8(5), 366–368.
36. Sheehan, J. E., Buesking, K., & Sullivan, B., (1994). Carbon-carbon composites. Annual Review of Materials Science, 24(1), 19–44.

37. Vogel, L., Wonner, P., & Huber, S. M., (2019). Chalcogen bonding: An overview. Angewandte Chemie International Edition, 58(7), 1880–1891.
38. Wadt, W. R., & Hay, P. J., (1985). Ab initio effective core potentials for molecular calculations. Potentials for main group elements Na to Bi. The Journal of Chemical Physics, 82(1), 284–298.
39. Weetman, C., & Inoue, S., (2018). The road travelled: After main□ group elements as transition metals. ChemCatChem, 10(19), 4213–4228.
40. Wiberg, N., (1997). Sterically overloaded supersilylated main group elements and main group element clusters. Coordination Chemistry Reviews, 163, 217–252.
41. Windhorst, T., & Blount, G., (1997). Carbon-carbon composites: A summary of recent developments and applications. Materials & Design, 18(1), 11–15.
42. Zannoni, D., Borsetti, F., Harrison, J. J., & Turner, R. J., (2007). The bacterial response to the chalcogen metalloids Se and Te. Advances in Microbial Physiology, 53, 1–312.

CHAPTER 8

Symmetry as the Organizing Principle for Crystal Structures

Contents

Unit Introduction 194
8.1. Crystallographic Group–Subgroup Relations 196
8.2. The Symmetry Principle in Crystal Chemistry 199
8.3. Structural Relationships By Group–Subgroup Relations 200
8.4. Diamond–Zinc Blend 202
8.5. The Relation Between Nias and MNP 203
8.6. Symmetry Relations at Phase Transitions: Twinned Crystals 207
8.7. Summary 211
Review Questions: 212
References 213

UNIT INTRODUCTION

Dense packings of hard particles have been effective models for studying the structure of low-temperature states of matter, including liquids, glasses, crystals, granular media, diverse materials, and biological systems (such as tissue structure, cell membranes, and phyllotaxis). Since most of our information on dense packings pertains to sphere-shaped particles, it is helpful to briefly review the geometrical characteristics of sphere packings to put our findings on dense packings of non-sphere-shaped particles into proper perspective. The geometric-structure classification highlights the wide range of jammed sphere packing types that can be achieved with unpredictable enormities in general order (as described in a scalar order metric ψ that exist in the states), packing fraction symbolize as φ, and other broad parameters (for example, mean-average contact number) (Percec et al., 2021).

It is possible to identify and define the recognized packing particles, such as the heaviest sphere packing (Kepler's conjecture) and efficiently arbitrarily jammed (MRJ) packing, as extremely states among order map for a specific jamming category through the combination of the concept of "order maps" via mathematically precise "jamming categories." Along with these exceptional jammed states, a few of which were just recently identified as physically relevant, the picture also includes the jamming-threshold states (least compact jammed packings) and extreme states between these and the MRJ point. Recently, dense packages of ordered and disordered non-spherical entities have not attracted as much scholarly interest compared to dense packings of sphere-shaped, a simple form which does not tile Euclidean space (Torquato & Jiao, 2012).

The densest packing of corresponding particles d have spherical, such as both shapes like concave or convex, are the subject of the present chapter. We will demonstrate that the maximal-density configurations are determined by the coupling of rotational units of freedom through the translational units of freedom, which depends critically on the particle surface's regularity and local primary curves of symmetry. As a result, various shapes will have maximum configurations of particle density with various kinds of packing properties.

It's interesting to note that the prediction for a few radially symmetrical polyhedra has just lately undergone experimental confirmation. Concave areas along with additional convex particles are added to these organizing ideas. Principles of concave shapes have already been expressed directly and implicitly (without more justification). Planning of different innovative crystals, glass or liquid matter states which are considerably more complicated than those made feasible by spheres is now achievable – thanks to particle shape tweaking (Gershoni & Kobayashi, 2005). In spite of the fact that the aforementioned is noticeable – that solid spheres entropically caused disorder sequence changes, metastable disorganized states, glassy jammed states, as well as directed jammed states with maximum density, the corresponding phase diagram for diligently non-spherical atoms will typically be significantly more complex due to the rotation angles of independence and uniformity of the surface of the particle. It is frequently brought up how polyhedral building blocks are packed, for instance in relation to the good distinction amongst the cadmium chloride types like $CdCl_2$ and CdI_2 (Ockwig, Delgado-Friedrichs, O'Keeffe, & Yaghi, 2005).

Together, all of them are made up of the same edge-sharing octahedra that overlap considerably. When each of the layer's stacked this way, the halogen atoms produce a close hexagonal packing for the cadmium CdI_2 type, while the other one is a cubic closest-packing (such as $CdCl_2$) type. The octahedral spaces inside particle packing are occupied by the metal atoms. The connection between that of a polyhedral and their related chemical properties are the key topics of interest, whereas the arrangement of the ions or molecules in the structure of crystals is considered as a secondary element. However, we will discover comparable findings from the standpoint of general packing and the most vital packing value, the closest packing sphere-shaped, is the main topic of discussion (Katzir, 2004).

The same idea that is applicable to halides of the many cadmium elements also applies to many other compounds: when the atomic particles are taken individually, a fraction of the compounds form the nearby packing, and the remainder of atoms reside in the empty spaces left in this packing. In order to characterize compounds with tightly packed atoms, the atoms that make up the packing do not all have to be of the same type or size. For instance, the titanium elements in perovskite, $CaTiO_3$, fill some octahedral interstices like that of the Ca and O_2 molecules form a well-defined cubic packing (Ermolenko, 2005).

Due to the size of the atoms in the spaces and their mutually reinforcing interaction with the neighboring atoms, there may be some distortions in the spherical packing, but these are typically minimal. In addition, if the packing is made larger, it might be possible to include atoms that are too big to fit into the interstices; in this situation, the packing no longer meets the criteria for being a closest-packing because its spheres lack any interaction with each other, but the relative position of the atoms is still possible in theory (Gershoni & Kobayashi, 2005).

Did You Know?

The components of symmetry existing in a specific crystalline solid govern its form and affect its physical characteristics.

8.1. CRYSTALLOGRAPHIC GROUP–SUBGROUP RELATIONS

A set of operations about symmetry which consistently satisfies the mathematical definition of a group are called a "space group." A powerful, accessible mathematical tool for ordering a wide range of crystalized structures countenancing to their space groups is called the "group theory." Without looking into great detail regarding group theory, we provide a few concepts in order to accomplish this. The information that follows isn't strictly mathematical; rather, it's an explanation of the facts with examples. For our purposes, this does not result in any errors. All space groups are infinite, which makes the precise mathematical treatment challenging, a precise treatment (Ivantchev, Kroumova, Madariaga, Perez-Mato, & Aroyo, 2000).

There are multiple symmetry operations in the space group G_1. If a space group G_2 includes some of these symmetrical processes, then it is a subset of G_1, and vice versa. G_1 and G_2 are supergroups. Symmetrical group operations within the space group as included in the G_1 which is multiplying an atom at an ordinary position by a factor of n_1. By analogy, for each matching atom, an atom belonging to subgroup G_2 is divided by n_2. Because G_2 has more symmetry operations than G_1, $n_1 > n_2$ is true. The proportion n_1/n_2 represents the degree of symmetry loss from G_1 to G_2. It is always an integer and is used to organize space groupings hierarchically. For instance, there is a reduction in index 3 symmetry when transitioning from rutile to tri rutile (Koch, 1984).

There are various subgroups which is the G_2, a maximum group in G_1 if there is no space group that might act as a connection assembly in G_1 and G_2 subgroups. Consequently, G_1 is a sub-group that is always smaller than G_2. Always, an index of symmetric decrease from a group to the maximum subgroup is a number that is prime or a unique number power.

The theorem of Hermann (1929) states that a maximum subgroup is likely the translationengleiche or a klassengleiche subgroup (Müller, 2011). When a subgroup is translationengleiche, it keeps its original translation lattice, which means the translation vectors' primitive unit cell sizes are the same for both the entire group and the subgroup. Subgroups that are translationengleiche maintain their original translation lattice, meaning the primitive unit cells' sizes for both the group and the subgroup's translation vectors are identical. In this instance, the absence of other symmetry processes, such as the reduction of the number of symmetry axes, is used to achieve symmetry reduction. This suggests that there will be a change in the crystal class (Mois et al., 2011).

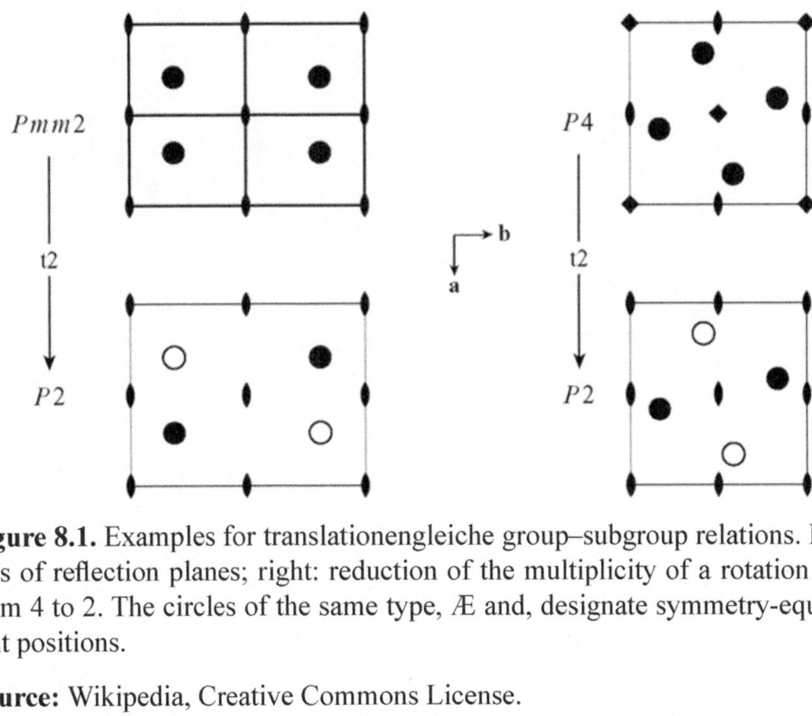

Figure 8.1. Examples for translationengleiche group–subgroup relations. Left: loss of reflection planes; right: reduction of the multiplicity of a rotation axis from 4 to 2. The circles of the same type, Æ and, designate symmetry-equivalent positions.

Source: Wikipedia, Creative Commons License.

The four-fold rotation axis in the example on the right of Figure 8.1 turns a dual rotational axis when about four symmetrically comparable atomic substances are substituted for double sets of diverse atoms; the interpretation vectors remain unaffected. Space group members and a klassengleiche subgroup belong to a similar crystal class. Reduced symmetry results from an absence of translational symmetry, that can be brought on by either enlarging a single cell or lost its centering. The quantity of other symmetry-related operations is also decreased as a result of the absence of translation symmetry (Elcoro et al., 2017). Two examples are shown in Figure 8.2. The illustration on the right side of Figure 8.2 demonstrates another way that increasing the unit cell size can result in a reflection plane symmetry being reduced to that of a glide plane. Similarly, when a rotational axis experiences cell enlargement along the axis, its symmetry can be decreased to that of a screw axis.

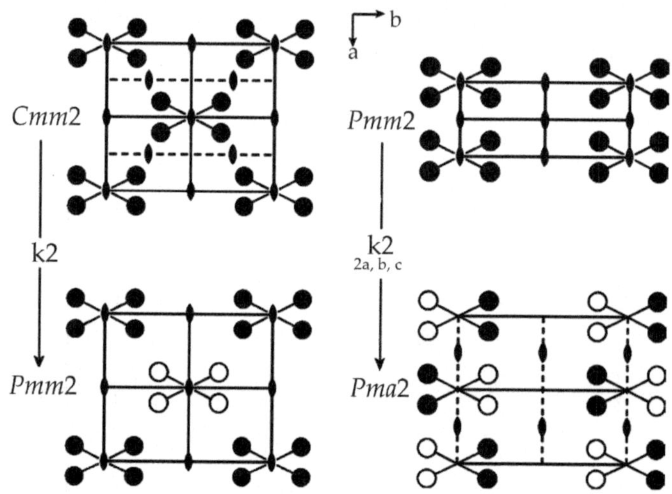

Figure 8.2. Examples of group-subgroup relationships that are klassengleiche. Right: Enlargement of the unit cell with symmetrical reduction of reflection planes perpendicular to b to glide planes. Left: Loss of centering, including loss of glide planes and twofold axes. Similar-type circles, such as and, designate symmetry-equivalent places.

Source: Wikipedia, Creative Commons License.

A specific example of isomorphic subgroup is the klassengleiche subgroup. Enantiomorphic space groups are the same thing as a group and a subgroup. They consequently have a Hermann-Mauguin symbol in common and an enantiomorphic space-based formation symbol (like $P3_1$ and $P3_2$). These main subgroups consist of a larger unit cell. Rutile and tri rutile are two examples. A well-suitable technique to display various subgroup links is with "family trees," showing the close ties between the space groups and largest subgroups by arrows indicating downward (Mois et al., 2006).

Each arrow's center is marked with the relationship type while the symmetry reduction index: t2 for translationengleiche subset of index 2, k2 for klassengleiche subgroups of index 2, and i3 for isomorphic subgroups of index 3. A new unit cell developed from the old one is also described, if necessary (see Figure 8.2 for the subgroup basis vectors may be presented as the vectorial sum of the base vectors of the prior group) (Müller & Wondratschek, 2005). There are times when a beginning change must also be declared. This is the case when the requirements of the International Table for Crystallography rules require that the source of the subgroups can be placed in a changed position from the source of the prior space. A number triplet represents the origin shift, such as 0, 1/2, –1/4 = shift by 0, ½ b, –¼ c, where b and c are the base vectors of the previous space group (Poettgen, 2014).

Remember

Geometry, nature of particles, and shapes are all fundamentally symmetrical. It produces patterns that aid in theoretically organizing our reality. We regularly miss regularity even though we see it all the time. In their certified lives, people relate symmetry ideas like rotations, tessellations, translations and reflections.

8.2. THE SYMMETRY PRINCIPLE IN CRYSTAL CHEMISTRY

High symmetry arrangements tend to form in crystalline solids, as can be seen. The symmetry principle, proposed by Laves (1950) in this form, has been stated more precisely by Barnighausen (1985).

- Atoms in the solid-state condition exhibits a noticeable shift towards the maximum symmetry.
- Contradictory characteristics of the atoms or atoms may make achieving the highest degree of symmetry impossible, but these deviations are typically minor.
- By virtue of the alignment of domains made inside the crystal-like matrix, the developed symmetry of the preliminary substance is often ultimately maintained during a phase change or reactions in a solid-state that revenues one or lower proportion symmetrical products (Müller, 2013).

Similar atoms typically occupy equivalent positions. The following formulation sheds light on the physical foundation of the symmetry concept: For example, in each type of atom, there is at least one dynamically most advantageous spatial organization that all atoms tend to take on, depending on a number of factors, including pressure, temperature, relative size of atoms, types of chemical bonding, electronic configuration and chemical composition, etc. According to quantum mechanics, particles are indistinguishable from atoms (Belokoneva, 2005).

They are, however, only indistinguishable in a crystal if their symmetry is equivalent because only then do they share the same environment. Further, use of close packing of spheres as an illustration of how common high-symmetry structures are by noting that only the cubic-shaped and hexagonal spheres packing contain all those atoms which have equivalent symmetry; other variations of closed packing contain various nonequivalent atomic positions, and these close packing are relatively infrequent. The equivalent places for all atoms are not always possible under the given conditions. Consider the following instances as an illustration: MX_5 composition; all X atoms are covalently bonded to M atoms (Pannetier, Bassas-Alsina, Rodriguez-Carvajal, & Caignaert, 1990).

Only if each group of five X atoms forms a standard pentagon surrounding an M atom (as in the instance of the XeF_5 ion, for example) can all the X atoms in this scenario be equal. The X atoms must be in at least two non-equivalent places if for any reason this is not achievable. The symmetry principle states that there should be a minimum number of non-equivalent places (Filatov, 2011).

8.3. STRUCTURAL RELATIONSHIPS BY GROUP–SUBGROUP RELATIONS

Numerous solid structures can be understood as products or derivative of simple, structure types with higher symmetry. Let's consider a few instances:

- Body-centered cubic sphere packing superstructures of CsCl CsCl type
- Diamond zinc blende Chalcopyrite
- Closest-packings closest-packings with occupied octahedral interstice (Hoffmann & Pöttgen, 2001).

Every time, we begin with a straightforward structure with high symmetry. Each arrow () in the examples above denotes a group-subgroup relationship, or a reduction of symmetry. These are perfect tools for systematically revealing structural relationships because they are mathematically well-defined (Müller, 2011). Changes like the following could be the cause of symmetry reductions:

- Different types of atoms replace the symmetry-equivalent

positions of an element's atoms. CC (diamond) ZnS (zinc blende), for instance.

- Voids either take the place of atoms or are surrounded by atoms. Consider the CdI_2 type of hexagonal closest-packing. This can be interpreted as a "substitution" of gaps by atoms if the empty spaces are thought of as "zero atoms."
- Atoms from one element are swapped out for atoms from another that calls for a different kind of bonding. Consider the perovskite type material $KMgF_3$ $CsGeCl_3$(such as one electron pair at the Ge atom; also, the Ge atom moved from the center of the octahedron to one of its faces to form the three covalent bonds that make up a $GeCl_3$ ion).
- Errors brought on by the Jan-Teller effect. For instance: $CuBr_2$ (distorted CdI2 type) $CdBr_2$ (CdI_2 type).
- The appearance of fresh interactions. For instance: I_2 molecules (normal pressure) – iodine (strong pressure, metallic, packing of spheres).
- Errors brought on by covalent bonds. For instance, $RuCl_3$ (low temperature, orthorhombic, Ru-Ru bonds) and $RuCl_3$ (high temperature, hexagonal TiI_3 type).
- Phase changes. Examples include such as the $CaCl_2$ (>217°C, rutile type) $CaCl_2$ (>217°C), and $BaTiO_3$ (>120°C, cubic perovskite form) $BaTiO_3$ (<120°C, tetragonal), Figure 8.5. A group or subgroup relationship among the space groups is necessary for second-order phase transitions (Hornsey & Hogg, 2000).

The Barnighausen-proposed family tree structure of group-subgroup relations can be used to depict structural relations in a clear and succinct way. The International Space Group Tables for Crystallography, Volume A1 can be used to set them up, and it has a thorough list of the concentrated subgroups of each space groups. The aristotype (or basic structure) is the framework with the highest degree of symmetry and is situated at the highest point of the familial tree (Barnighausen tree), from which every other structure in an organized family is descended. Hettotypes is the structure (or derivation structures) that is produced after the reduction of atomic symmetry (Bendersky, Roytburd, & Boettinger, 1994).

Each space group listed in a family tree has a structure. The space group symbol itself just conveys symmetry and gives out nothing about the positions of the atoms, therefore extra information for each member of the "family tree" is needed, including the Wyckoff symbol, site anatomic as well as symmetrical categories. The informational worth of a family tree is quite limited in its lack of these facts. In basic circumstances, the data may be added to the tree, but in complex cases, an independent table is more useful. The examples exhibit how to build site occupation specifications (Belkyal, Mokhlisse, Tanouti, Hesse, & Depmeier, 2000).

For the reason, they remain more illustrative, it is advised to designate the space groups using the full Hermann Mauguin symbols. It is possible to see that symmetry gradually decreases by looking at the atomic locations. An aristotype typically has all of its atoms in distinct places, such as on specific symmetry elements, with predetermined coordinate values, and they have specific site symmetries (Kahlenberg, Fischer, & Baur, 2001). From group to group and subgroup to subgroup, each atom location undergoes at least one of the alterations listed below:

- Site of symmetry is reduced. The simultaneous independence of the variables X, Y, and Z coordinate numbers may enable the atom to migrate away from repaired values of a particular place.
- The partition of a set of independent, separate positions into a set of symmetry-equivalent locations (Girgsdies, Schneider, Brückner, Ressler, & Schlögl, 2009).

8.4. DIAMOND–ZINC BLEND

Reduction of zinc blende symmetry to the diamond group-subgroup connection. In both structures, the atomic positions and site imbalance are identical. In the diamond unit cell, there are about eight Carbon atoms situated in symmetry corresponding locations (as in Wyckoff position 8a) (Kobashi et al., 1990). The reduction in symmetry caused the locations of atomic particles to split into two different positions for example 4a and 4c, which are then occupied by zinc and sulfur atoms in the zinc mix. Because they are translationengleiche, the unit cell sizes in the space groups coincide. Indicating that precisely one-half of all symmetry-related processes are lost and the degree of symmetry reduction index equals to 2. This has to do with the diamond's C-C bond inversion centers (Biernacki & Scheffler, 1989).

These symmetrical changes can be diminished further. A minimal subset of F 43 m is I 42d with a double lattice-framework parameter of c. Using the technique F 43 I 42d, the Wyckoff, who give the position to the particles of zinc divides once again, making it usable through using atoms of two different elements. Another F 43 meters tetragonal group with a doubled c axis is called I 42m. It separates locations 2a, 2b, and 4d from the zinc blende position 4c. Similar to the $FeSnCu_2S_4$ structure of stannite, they are filled with atoms of three distinct elements (Volodin & Sachkov, 2021).

8.5. THE RELATION BETWEEN NIAS AND MNP

To enable the replacement of the C atoms with atoms from other elements, the symmetry must be reduced to the aforementioned hettotypes of diamonds. Site symmetries must be reduced rather than Wyckoff positions being split to account for structural distortions. Let's go back to thinking of MnP as a warped variation of the nickel arsenide form. Figure 8.4 displays the relationships as well as pictures of the structures. In the first stage, the hexagonal symmetry must be absent; a minor deformation in the lattice framework will do (Selte, Kjekshus, & Andresen, 1973).

Typically, the orthorhombic in subgroups has to have a C-centered cell. Although the cell's size is doubled, centering makes it a translationengleiche subgroup. The appropriate cell change is declared in the center of the group or subgroups arrow. In the second stage, the centering is removed, which reduces the translations by 50%. As a consequence, the reduction of index 2 is a symmetry reduction with klassengleiche properties. The images in Figure 8.3 illustrate where symmetry components are deleted throughout the two stages of the symmetry drop. The removal of half each of the inversion centers is one aspect of the subsequent stage (Shen et al., 2023).

On the contrary, focuses of the space subgroups group $C2/m2/c2_1/m$ (also referred to as Cmcm) that have been eliminated are the "Wyckoff" coordinates such as 4a (0,0,0) and 4b (1/2,0,0), however, inversion centers of the "Wyckoff" location 8d (1/4, 1/4,0) have been kept. Because it is believed that the origin would be situated in an inversion centre, as shown by the group-subgroup arrow, the starting location of the subgroups $P2/m2_1/c2_1/n$ (short Pmcn) needs to be displaced by –1/4, –1/4, 0. This shift demands that the coordinates be modified by ¼, ¼, 0. Among the boxes with atom coordination, the necessary coordinate transformations are offered (Tremel, Hoffmann, & Silvestre, 1986).

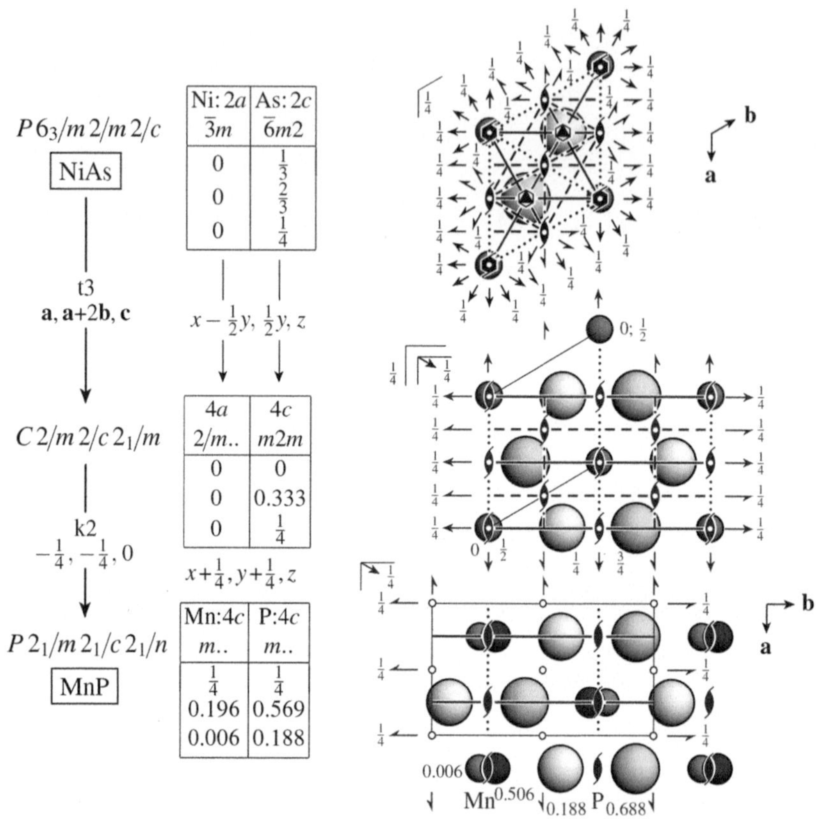

Figure 8.3. The connection between MnP and NiAs. The numbers represent z coordinates for the photos.

Source: Wikipedia, Creative Commons License.

By adding ¼, ¼, 0 to the values for the space group Cm presented in Figure 8.3, we can obtain ideal values for an unmodified structure in Pm cm. But because the distortion is absent, the symmetry stays at Cm cm. Moving the atoms from their optimum places is the only way to create a space group Cm cm. The Manganese atom y number (0.214 rather than 1/4) and the proton atom z number (0.207 rather than 1/4) are the initial deviations. Because these differences are so small, we have good reason to think that MnP is a deformed variant of the NiAs type. This is a formal representation of the link between zinc blende and diamond. It is not possible to substitute sulfur and zinc atoms for carbon atoms. However, it is conceivable to carry out Figure 8.3 NiAs structural deformation. This happens as a result of phase transitions. This kind of phase transition, for instance, happens in MnAs above 125°C (second-order phase change in NiAs at 125°C; another transition happens at 45°C) (Franzen, Haas, & Jellinek, 1974).

The octahedral spaces are filled when the hexagonal closest-packing occurs. The discussion states that a variety of organizations be able to be made using the hexagon closest-packing of the sphere by putting additional atoms inside certain of the octahedral interstices. If the X atoms from a compound MX_n create the sphere packing inside elements, the portion 1n of the octahedral spaces must be filled. As the unit cell has two octahedral interstices, only fractions of 1/1 or 1/2 of the octahedral space can be taken without enlarging the 3c) hexagonal closest-packing. The unit cell must be expanded to accommodate any further fractions (Franzen & Wiegers, 1975).

In other words, the only combinations that allow the creation of structures without cell enlargement are MX and MX_2. When a cell expands, translational symmetry is lost, which calls for the existence of klassengleiche group-subgroup connections. The NiAs type, which corresponds to the wrap in which all of the octahedral spaces are filled with nickel atoms (Wyckoff positioning 2a), or the packing of spheres themselves, can be considered as the aristotype. In the aristotype, these interstices have comparable symmetry, but if they are not filled with one kind of atom or another (or if the Ni atom in NiAs has been partially eliminated or replaced), a subgroup can arise (Motizuki, Katoh, & Yanase, 1986).

After this operation, the interstitial sites are no longer matched. There is just one maximum subgroup, P 32/m1, in the space group $P6_3/m2/m2/c$, each hexagon closest-packing of spheres, where 2a is divided into two autonomous places. If only one of these seats is filled and the other is open, the CdI_2 type is shown. The lattice unit cell of the hexagonal closed packings can remain tripled, as seen in Figure 8.4. The larger cell keeps its hexagonal or triangular shape if its center vector (viewing direction) isn't modified. A rhombohedral-shaped cell develops when c is also tripled and centers in 2/3, 1/3, 1/3, and 1/3, 2/3, 2/3 (Rundqvist et al., 1962).

Even though the cell's volume has increased by a factor of nine, the resulting primitive unit cell is only three times bigger because of the centrifugal effect. In both instances, there are six atomic spheres (X atoms) and six primitive octahedral holes in large cells. The occupied position of two of the octahedral holes—leaving the other four empty—leads to the composition M_2X_6 or MX_3. Using the aforementioned triplicated cells, a hexagonal closest-packing of spheres can be used to derive the structural features of some compounds, such as MX_3 and M_2X_3 (Kallel, Boller, & Bertaut, 1974).

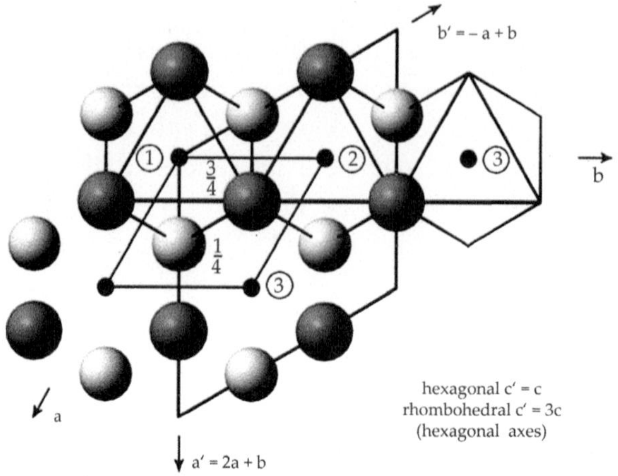

Figure 8.4. Sections of the hexagonal closest-packing. Gray area: Unit cell, space group P 63 m 2 m 2 c. Large cell: Base of the triple cell. The z coordinates of the spheres refer to c = c. The dots labeled ①, ②, and ③ mark six octahedral voids at z = 0 and z = 1 (for c = c) and at z = 0 and z = 1 (c = 3c).

Source: Wikipedia, Creative Commons License.

Two or six tiny boxes are utilized to symbolize the octahedral positions for every space group instead of assigning numerical values. The image upper-left corner displays the relevant coordinates. The same letter is used to identify the Wyckof (shown in Figure 8.4) symmetry-equivalent octahedral structures that correlate to the locations of the octahedron centers. The increasing diversity of letters reveals the decreasing regularity from top to bottom (Suzuki & Ido, 1982).

The right category of the family tree has a list of rhombohedral subgroups. The main unit cell is tripled at klassengleiche elimination P312/c—k3 R32/c. Moreover, there are two octahedral gaps in the location 2b of P312/c results to produce six octahedral voids in locations 2b and 4c of R32/c. Position 4c is vacant because the RhF_3 type relates to the occupancy of position 2b. Corundum (–Al_2O_3) is the reverse, whereby 4c is occupied but 2b is left unfilled. In the following phase,

R32/c t2 R3 further divides the places. BiI_3, ilmenite ($FeTiO_3$), $LiSbF_6$, and WCI_6, all possess various structural shapes depending on whether one is occupied (Johnson, 1975).

8.6. SYMMETRY RELATIONS AT PHASE TRANSITIONS: TWINNED CRYSTALS

By way mentioned before, reconstructive phases are always 1st order transitions. The transition begins at the nucleation site which could stand a spot where there is a void in the crystal. There, the nucleus grows, and the structure is restored, causing the atoms to start moving. As atoms perform the required diffusion, bonds from the old structure disintegrate at the boundary among the two phases and are replaced by new bonds from the emerging new structure (Pitteri & Zanzotto, 2002).

Interactions within the space groups group-subgroups are irrelevant in this kind of phase transition. Contrary to sporadic assumptions, it is unbearable to allocate an interplanetary group to the arrangement of the atoms at a phase boundary. These hypotheses suggest that first-order transitions between phases occur via a halfway or the prevalent subcategory of the atomic space groups involved. The scenario is different when the 2nd order phase transitions take place. These can individually occur when the space assembly of one stage is a (possibly small) subgroup of the space subgroup of the other phase. Most modifications to displacive structures are the 2nd order, where atomic sets only undergo very modest reciprocal motions (Darlington & Megaw, 1973).

As mentioned in the following example, the phase transition in the second-order members of the mineral known as the calcium chloride through metallic type in which $CaCl_2$ type is shown. The only action is a shared rotation of the octahedral strands of the edge type. As a result, it can be inferred that tetragonal symmetry is "broken" and cannot be preserved. A required subgroup of the space group corresponding to the $CaCl_2$ type is the space group of the rutile type $P4_2/m2_1/n2/m$. Transitioning from the tetragonal to the orthorhombic crystal class produces the translationengleiche subgroup. It demonstrates the correspondence. The lattice-framework parameters a and b shift away from a tetragonal shape to an orthorhombic form. Either a>b or a<b is the result, subject to how octahedral strands are rotated (Hazen & Finger, 1979).

Both rotating directions occur at random throughout the phase transition, resulting in the emergence of regions that are both a>b and a<b. A twin resulting crystal is composed of intergrown domains. The symmetry operation that connects the two identical domains and was lessen through the proportion decrease outlined in 3 element symmetric value corresponds to the mutual orientations of the twin domains (Wang & Lu, 2006).

***Definition*:** Twins are made when two or more large crystals of the same species grow together in a way that follows crystallographic rules. Other terms associated to it are twin domains, twin partners, or twin components (Gao, 2020).

It is very typical to see twinned crystals. They might consist of dissimilar actualities that can be visible in macroscopic instruments, like the mirror-inverted "dovetail twins" of gypsum (as shown in Figure 8.5). There could be an abundance of many alternate components, which can occasionally give crystal an inconsistent appearance (polysynthetic twin). As in the example of dovetail twins, a symmetry operation (in some cases referred to as a "twinning operator") converts one of the twin parts into the other. Another instance includes the interconversion of the quartzite "Dauphin'e twins," that are split apart by a double-rotating axis (see Figure 8.5) (Ali, Nuss, Sheptyakov, & Jansen, 2010).

Threefold or fourfold axis, which gives domains either three or four orientations, can also exist as symmetry components among the components. Even if this pairing operation is not a symmetrical functioning among structures of space group, it must agree with the given structural facts. Twin creation is anticipated when there is a group-subgroup relation and a phase transition from one fewer symmetric space group to another (Kiho, Peterlin, & Geil, 1964).

If the translationengleiche subgroup has an index of 2, three, or four, the twin will have two types of domains, three types, and four types, respectively (indices greater than 4 are not discovered between translationengleiche maximal-subgroups). If a symmetrical reduction takes several stages and there are two translationengleiche processes, twins of twins may appear. In high-temperature modifications, phase transitions caused by temperature changes often have greater amounts of symmetry. These Dauphin'e twins are formed as quartz change due to the high-temperature form, also known as high quartz, to its low-temperature form, which is stable above 573°C (Ellner, 1995).

These space groups are also known as quartz, a translationengleiche subgroup that has second indices of the quartz space group, is responsible for producing twins. The final two twofold axes in the blank space group symbols $P6_222$ disappear during the symmetry reduction. The pairing operation, which only exists in the one with a greater symmetry space group, is produced by these axes. If they were present, they would connect an atom from the position x y z through the sites x, x–y, 2/3 –z, y–x, y, 1/3–z, as well as –y, –x, –z in the coordinate system of P_321 (Herbstein, 1965). Contrast this to the atomic coordinates of both of the twin components in Figure 8.6.

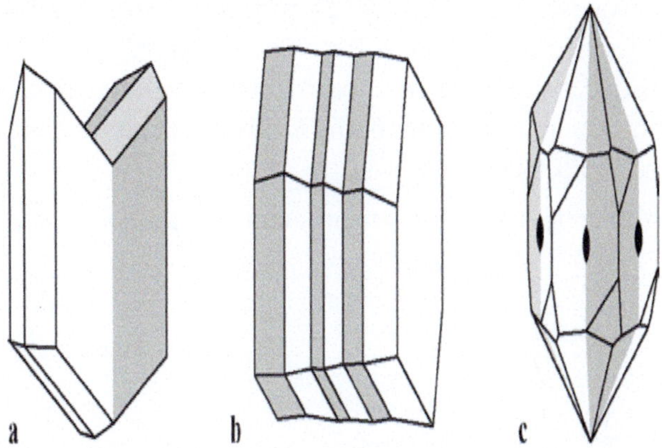

Figure 8.5. Group–subgroup relations and the emergence of twins in the phase transition β-quartz α-quartz (Dauphin'e twins).

Source: Wikipedia, Creative Commons License.

Did You Know?

Balance in strategy produces synchronization, order, and consequences that are visually pleasant because symmetrical objects promote balance. Nature contains it in great quantity, which is possibly what we consider it to be so appealing.

These quartz crystals cannot be used as piezoelectric constituents for electrical strategies because the polar instructions that constitute the twins compensate for one another. Therefore, when creating piezoelectric quartz, temperatures must never exceed up to the 573°C. The reduction in the symmetry as shown in Figure 8.4 produced the space group known as $RuBr_3$ ($P2_1/m2/n2_1/m$), which has a translationengleiche phases with index 3. The structure just slightly departs from the hexagon TiI_3 type by moving the elements such as Ru from the octahedron centers, and at warmer temperatures, the greater symmetry is kept. $RuBr_3$ that was made at greater temperatures, then refrigerated features twinned crystals containing three different elements that are independently rotated by 120° as a result of the removal of the tripartite structure at step $P6_3/m2/2m$ t3 $C2/m2/c2_1/1m$ (Kim, Nettleship, & Kriven, 1992).

As a consequence of the X-ray diffraction research, where the X-ray components are overlaid and provide the pattern hexagonal symmetry, the Ru atom placements seems to be precisely in the centers of the octahedrons.

Non-identified twins that appear to have incorrect symmetry might provide a difficult difficulty and lead to false conclusions throughout the framework determination process. Twins can develop not just during solid-state phase transitions but also during crystal formation. The intergrowth of developing twins is controlled by the emergence of crystallization nuclei. Relations between distinct groups and their subdivisions are irrelevant in this situation (Conti & Zanzotto, 2004) (Table 8.1).

Table 8.1. Crystal data of structures mentioned in Figure 8.4

Space group	Compound	*a* /pm	*c* /pm	Wyckoff position	Element	*x*	*y*	*z*	Ideal coordinates *x*	*y*	*z*
$P6_3/mcm$	TiI_3 hex.	714.2	651.0	2*b*	Ti	0	0	0	0	0	0
				6*g*	I	0.683	0	$\frac{1}{4}$	0.667	0	$\frac{1}{4}$
$P\bar{3}1m$	Ag_3O	531.8	495.1	2*c*	O	$\frac{1}{3}$	$\frac{2}{3}$	0	$\frac{1}{3}$	$\frac{2}{3}$	0
				6*k*	Ag	0.699	0	0.277	0.667	0	0.25
	Li_2ZrF_6	497.3	465.8	1*a*	Zr	0	0	0	0	0	0
				2*d*	Li	$\frac{1}{3}$	$\frac{2}{3}$	$\frac{1}{2}$	$\frac{1}{3}$	$\frac{2}{3}$	$\frac{1}{2}$
				6*k*	F	0.672	0	0.255	0.667	0	0.25
$P\bar{6}c2$	$LiScI_3$	728.6	676.8	2*a*	Sc	0	0	0	0	0	0
				2*c*	Li	$\frac{1}{3}$	$\frac{2}{3}$	0	$\frac{1}{3}$	$\frac{2}{3}$	0
				6*k*	I	0.673	−0.004	$\frac{1}{4}$	0.667	0	$\frac{1}{4}$
$P6_322$	Ni_3N	462.2	430.6	2*c*	N	$\frac{1}{3}$	$\frac{2}{3}$	$\frac{1}{4}$	$\frac{1}{3}$	$\frac{2}{3}$	$\frac{1}{4}$
				6*g*	Ni	0.328	0	0	0.333	0	0
$P\bar{3}1c$	$FeZrCl_6$	628.4	1178.8	2*a*	Fe	0	0	$\frac{1}{4}$	0	0	$\frac{1}{4}$
				2*c*	Zr	$\frac{1}{3}$	$\frac{2}{3}$	$\frac{1}{4}$	$\frac{1}{3}$	$\frac{2}{3}$	$\frac{1}{4}$
				12*i*	Cl	0.667	−0.023	0.131	0.667	0.0	0.125
	$LiCaAlF_6$	500.8	964.3	2*b*	Ca	0	0	0	0	0	0
				2*c*	Al	$\frac{1}{3}$	$\frac{2}{3}$	$\frac{1}{4}$	$\frac{1}{3}$	$\frac{2}{3}$	$\frac{1}{4}$
				2*d*	Li	$\frac{2}{3}$	$\frac{1}{3}$	$\frac{1}{4}$	$\frac{2}{3}$	$\frac{1}{3}$	$\frac{1}{4}$
				12*i*	F	0.623	−0.031	0.143	0.667	0.0	0.125
	$AgInP_2S_6$	618.2	1295.7	2*a*	In	0	0	$\frac{1}{4}$	0	0	$\frac{1}{4}$
				2*d*	Ag	$\frac{2}{3}$	$\frac{1}{3}$	$\frac{1}{4}$	$\frac{2}{3}$	$\frac{1}{3}$	$\frac{1}{4}$
				4*f*	P	$\frac{1}{3}$	$\frac{2}{3}$	0.164	see text		
				12*i*	S	0.657	−0.028	0.120	0.667	0.0	0.125

*The ideal coordinates would apply to an undistorted packing of spheres. Coordinate values fixed by symmetry are stated as 0 or fractional numbers, otherwise as decimal numbers.

Did You Know?

All crystals have three-dimensional in nature translational symmetry, but some have additional symmetry components. Since the crystal possesses twofold symmetry of rotation about this axis, turning it 180 degrees may produce an atomic arrangement completely identical to the initial configuration.

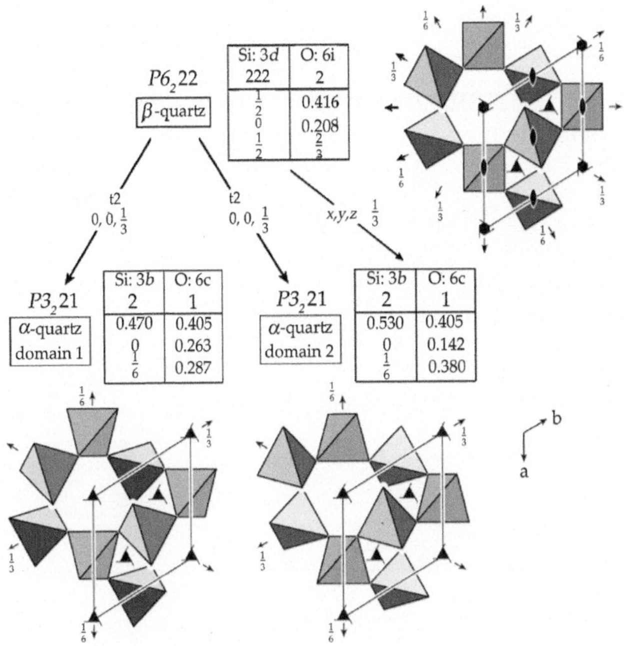

Figure 8.6. Group–subgroup relations and the emergence of twins in the phase transition β-quartz α-quartz (Dauphin'e twins).

Source: Wikipedia, Creative Commons License.

8.7. SUMMARY

This chapter has covered the symmetry principle of crystal structures. Crystallographic group and subgroup relationships are described. Group theory provides a powerful, mathematically simple tool for organizing a wide variety of crystal structures based on their space groups. Despite delving into great detail regarding group theory, we presented a few concepts to do this. A detailed explanation of the symmetrical concept in crystal chemistry, which was first postulated by F. Laves and later clarified by H. Barnighausen, is also provided. Group-subgroup relationships including the group-subgroup relations of the elements such as diamond and zinc blende symmetric reduction are described as having structural linkages. A specific example of klassengleiche subgroups is the isomorphic subgroups. Enantiomorphic space groups are the same thing as a group and a subgroup. As a result, they share a Hermann-Mauguin symbol or an enantiomorphic group of space-form symbol. It is discussed in detail that reconstructive phase shifts and symmetry relationships at phase transitions in twinned crystals are both first-order transitions. The change begins at the nucleation site, which might be a spot where there is a void in the crystal, and from there the atoms start moving, the nucleus grows, and the structure is rebuilt. The entire process is explained step by step.

REVIEW QUESTIONS:

1. What is crystallography?
2. Describe in detail the diamond zinc blende?
3. What are translationengleiche subgroups?
4. What are a klassengleiche subgroups?
5. What is hexagonal closest-packing?

REFERENCES

1. Ali, N. Z., Nuss, J., Sheptyakov, D., & Jansen, M., (2010). The $AFeO_2$ (A= K, Rb and Cs) family: A comparative study of structures and structural phase transitions. Journal of Solid-State Chemistry, 183(3), 752–759.
2. Aroyo, M. I., Perez-Mato, J. M., Capillas, C., Kroumova, E., Ivantchev, S., Madariaga, G., & Wondratschek, H., (2006). Bilbao crystallographic server: I. Databases and crystallographic computing programs. Zeitschrift für Kristallographie-Crystalline Materials, 221(1), 15–27.
3. Aroyo, M. I., Perez-Mato, J. M., Orobengoa, D., Tasci, E., de la Flor, G., & Kirov, A., (2011). Crystallography online: Bilbao crystallographic server. Bulg. Chem. Commun, 43(2), 183–197.
4. Bärnighausen, H., & Schiller, G., (1985). The crystal structure of A-Ce_2O_3. Journal of the Less Common Metals, 110(1–2), 385–390.
5. Belkyal, I., Mokhlisse, R., Tanouti, B., Hesse, K.-F., & Depmeier, W., (2000). Low-temperature phase transition and structural relationships of (CH_3NH_3) 3Bi2Cl9. Journal of Solid-State Chemistry, 155(2), 286–291.
6. Belokoneva, E. L., (2005). Borate crystal chemistry in terms of the extended OD theory: Topology and symmetry analysis. Crystallography Reviews, 11(3), 151–198.
7. Bendersky, L., Roytburd, A., & Boettinger, W., (1994). Phase transformations in the $(Ti, Al)_3$ Nb section of the Ti–Al–Nb system—I. Microstructural predictions based on a subgroup relation between phases. Acta metallurgica et materialia, 42(7), 2323–2335.
8. Biernacki, S., & Scheffler, M., (1989). Negative thermal expansion of diamond and zinc-blende semiconductors. Physical Review Letters, 63(3), 290.
9. Conti, S., & Zanzotto, G., (2004). A variational model for reconstructive phase transformations in crystals, and their relation to dislocations and plasticity. Archive for Rational Mechanics and Analysis, 173, 69–88.
10. Darlington, C., & Megaw, H. D., (1973). The low-temperature phase transition of sodium niobate and the structure of the low-temperature phase, N. Acta Crystallographica Section B: Structural Crystallography and Crystal Chemistry, 29(10), 2171–2185.
11. Elcoro, L., Bradlyn, B., Wang, Z., Vergniory, M. G., Cano, J., Felser, C., & Aroyo, M. I., (2017). Double crystallographic groups and their

representations on the Bilbao crystallographic Server. Journal of Applied Crystallography, 50(5), 1457–1477.

12. Ellner, M., (1995). Polymorphic phase transformation of Fe4Al13 causing multiple twinning with decagonal pseudo-symmetry. Acta Crystallographica Section B: Structural Science, 51(1), 31–36.
13. Ermolenko, A., (2005). Two-plane symmetry in the structural organization of man. Medical Hypotheses, 64(1), 209–214.
14. Filatov, S., (2011). General concept of increasing crystal symmetry with an increase in temperature. Crystallography Reports, 56, 953–961.
15. Franzen, H., & Wiegers, G., (1975). The NiAs–MnP phase transition in VS. Journal of Solid-State Chemistry, 13(1–2), 114–117.
16. Franzen, H., Haas, C., & Jellinek, F., (1974). Phase transitions between NiAs-and MnP-type phases. Physical Review B, 10(4), 1248.
17. Gao, Y., (2020). A Cayley graph description of the symmetry breaking associated with deformation and structural phase transitions in metallic materials. Materialia, 9, 100588.
18. Gershoni, S., & Kobayashi, H., (2005). Symmetry and complementarity-a discussion and a qualitative demonstration with computer generated infinite patterns of natural scenes. Journal of The Society of Photographic Science and Technology of Japan, 68(2), 89–90.
19. Girgsdies, F., Schneider, M., Brückner, A., Ressler, T., & Schlögl, R., (2009). The crystal structure of δ-VOPO4 and its relationship to ω-VOPO4. Solid State Sciences, 11(7), 1258–1264.
20. Hazen, R. M., & Finger, L. W., (1979). Polyhedral tilting: a common type of pure displacive phase transition and its relationship to analcite at high pressure. Phase Transitions: A Multinational Journal, 1(1), 1–22.
21. Herbstein, F., (1965). Twinned crystals. II. α-l,2:4,5-Tetrachlorobenzene. Acta Crystallographica, 18(6), 997–1000.
22. Hoffmann, R.-D., & Pöttgen, R., (2001). AlB2-related intermetallic compounds – a comprehensive view based on group-subgroup relations. Zeitschrift für Kristallographie-Crystalline Materials, 216(3), 127–145.
23. Hornsey, M. J., & Hogg, M. A., (2000). Assimilation and diversity: An integrative model of subgroup relations. Personality and Social Psychology Review, 4(2), 143–156.

24. Ivantchev, S., Kroumova, E., Madariaga, G., Perez-Mato, J., & Aroyo, M., (2000). SUBGROUPGRAPH: A computer program for analysis of group–subgroup relations between space groups. Journal of Applied Crystallography, 33(4), 1190–1191.
25. Johnson, V., (1975). Diffusionless orthorhombic to hexagonal transitions in ternary silicides and germanides. Inorganic Chemistry, 14(5), 1117–1120.
26. Kahlenberg, V., Fischer, R., & Baur, W., (2001). Symmetry and structural relationships among ABW-type materials. Zeitschrift für Kristallographie-Crystalline Materials, 216(9), 489–494.
27. Kallel, A., Boller, H., & Bertaut, E., (1974). Helimagnetism in MnP-type compounds: MnP, FeP, CrAs and CrAs1–xSbx mixed crystals. Journal of Physics and Chemistry of Solids, 35(9), 1139–1152.
28. Katzir, S., (2004). The emergence of the principle of symmetry in physics. Historical Studies in the Physical and Biological Sciences, 35(1), 35–65.
29. Kiho, H., Peterlin, A., & Geil, P., (1964). Polymer deformation. VI. Twinning and phase transformation of polyethylene single crystals as a function of stretching direction. Journal of Applied Physics, 35(5), 1599–1605.
30. Kim, Y. J., Nettleship, I., & Kriven, W. M., (1992). Phase transformations in dicalcium silicate: II, TEM studies of crystallography, microstructure, and mechanisms. Journal of the American Ceramic Society, 75(9), 2407–2419.
31. Kobashi, K., Inoue, T., Tachibana, H., Kumagai, K., Miyata, K., Nishimura, K., & Nakaue, A., (1990). Selected-area deposition of diamond films. The Physics and Chemistry of Carbides, Nitrides and Borides, 159–167.
32. Koch, E., (1984). The implications of normalizers on group–subgroup relations between space groups. Acta Crystallographica Section A: Foundations of Crystallography, 40(5), 593–600.
33. Laves, F., (1950). The lattice and twinning of microcline and other potash feldspars. The Journal of Geology, 58(5), 548–571.
34. Li, R., & Emsley, J., (2013). The organizing principle of the platelet glycoprotein Ib–IX–V complex. Journal of Thrombosis and Hemostasis, 11(4), 605–614.
35. Motizuki, K., Katoh, K., & Yanase, A., (1986). Electronic band

structures of NiAs-type compounds. I. Nonmagnetic state. Journal of Physics C: Solid State Physics, 19(4), 495.

36. Müller, U., & Wondratschek, H., (2005). International Tables for Crystallography: Volume A1 Symmetry Relations between Space Groups: Springer.
37. Müller, U., (2011). Relating Crystal Structures by Group–Subgroup Relations, 3(2), 3–5.
38. Müller, U., (2013). Symmetry Relationships between Crystal Structures: Applications of Crystallographic Group Theory in Crystal Chemistry (Vol. 18). OUP Oxford.
39. Ockwig, N. W., Delgado-Friedrichs, O., O'Keeffe, M., & Yaghi, O. M., (2005). Reticular chemistry: Occurrence and taxonomy of nets and grammar for the design of frameworks. Accounts of Chemical Research, 38(3), 176–182.
40. Pannetier, J., Bassas-Alsina, J., Rodriguez-Carvajal, J., & Caignaert, V., (1990). Prediction of crystal structures from crystal chemistry rules by simulated annealing. Nature, 346(6282), 343–345.
41. Percec, V., Wang, S., Huang, N., Partridge, B. E., Wang, X., Sahoo, D., & Jezorek, R. L., (2021). An accelerated modular-orthogonal Ni-catalyzed methodology to symmetric and nonsymmetric constitutional isomeric AB2 to AB9 dendrons exhibiting unprecedented self-organizing principles. Journal of the American Chemical Society, 143(42), 17724–17743.
42. Pitteri, M., & Zanzotto, G., (2002). Continuum Models for Phase Transitions and Twinning in Crystals. CRC Press.
43. Poettgen, R., (2014). Coloring, distortions, and puckering in selected intermetallic structures from the perspective of group□subgroup relations. Zeitschrift für Anorganische und Allgemeine Chemie, 640(5), 869–891.
44. Rundqvist, S., Sillén, L. G., Timm, D., Motzfeldt, K., Theander, O., & Flood, H., (1962). Phosphides of the B31 (MnP) structure type. Acta Chem. Scand, 16(2), 287–292.
45. Selte, K., Kjekshus, A., & Andresen, A., (1973). Phase transitions between the MnP and NiAs type structures. Acta Chemica Scandinavica, 27(9), 3195–3206.
46. Shen, Z., Cao, M., Wen, Y., Li, J., Zhang, X., Hui, J., & Zhang, H., (2023). Tuning the local coordination of CoP1–x S x between NiAs-

and MnP-type structures to catalyze lithium–sulfur batteries. ACS Nano, 17(3), 3143–3152.

47. Suzuki, T., & Ido, H., (1982). Relation between structural and magnetic properties of compound MnAs1-x P x (0≦ x≦ 0.275). Journal of the Physical Society of Japan, 51(10), 3149–3156.
48. Torquato, S., & Jiao, Y., (2012). Organizing principles for dense packings of nonspherical hard particles: Not all shapes are created equal. Physical Review E, 86(1), 011102.
49. Tremel, W., Hoffmann, R., & Silvestre, J., (1986). Transitions between NiAs and MnP type phases: An electronically driven distortion of triangular (36) nets. Journal of the American Chemical Society, 108(17), 5174–5187.
50. Volodin, V., & Sachkov, V., (2021). Anisotropy of optical phonons in biaxially stressed zinc-blende-and diamond-type semiconductors and alloys. Physica B: Condensed Matter, 614, 413008.
51. Wang, W., & Lu, H., (2006). Twinning induced by the rhombohedral to orthorhombic phase transition in lanthanum gallate ($LaGaO_3$). Physics and Chemistry of Minerals, 33(7), 435–444.

INDEX

Symbols

3-dimensional grid 119

A

Acidic liquid 184
Actual atomic 3
Additive values 30
Alcohols 89, 91, 93, 95, 112
Aldehyde functional group 92
Aldehydes 92
Alkanes 87, 90, 111
Alkenes 90, 111
Aluminum cations 127
Amines 93
Amino 93
Anisotropic crystal 148
Anisotropy 147, 149, 163
Anisotropy crystal 149
Antimony 4, 5
Aromatic alcohols 161
Aromatic benzene ring 94
Atmosphere 133, 134
Atom dispersion 173
Atomic arrangement 67
Atomic configuration 61
Atomic coordinates 8, 9, 18, 23, 24
Atomic nuclei 4
Atomic orbitals 96, 106
Atomic particles 60, 64, 175, 181
Atomic polarizabilities 154
Atomic systems 123
Atomization energy 63, 66, 68
Atoms 2, 3, 4, 5, 6, 8, 9, 10, 11, 13, 14, 15, 16, 17, 18, 24
Atoms adopt hybridizations 96
Atom's coordination number 6
Attraction-repulsion forces 99

B

Bending graphite 184
Body-centered cubic arrangement 180
Body-centered cubic pack 183
Boiling temperatures 154
Bond polarity 156, 162
Born-Haber cycle 12
Boron-rich borides 185, 187
Bromine chlorine 187
Buckminsterfullerene 183
Burns violently 126

C

Carbon atom 87, 92, 93, 99
Carbon-hydrogen bonds 116
Carbon modification 181, 187
Carbon Nanotubes 184, 190
Carbonyl groups 89
Cartesian 8
Cell volume 7
Chair conformation 174
Chemical bonds 11, 12, 18, 19, 21, 23, 24, 25
Chemical bonds releases energy 130
Chemical bond theory 11, 18
Chemical composition 61
Chemical formula 127
Chemical nomenclature 117
Chemical reaction 2
Chemical reactivity 88
Chemical structures 2
Chlorine atoms produce chlorine gas 126
Classic refraction 146
Closer-together atoms 4
Closest neighbor 4
Clustering 89
Complex situations 4
Conduction band 16, 17
Conjugate acids and bases 104
Connected polyhedra 6
Contemporary chemistry 2, 10
Contraventions 8
Conventional chemical models 181
Coordinate axes 8
Coordinate Covalent Bonding 130
Coordinate refractions 162
Coordinate system 8
Coordination polyhedron 4, 5, 69
Corresponding atom 8
Covalent 11, 18, 21, 22
Covalent bonding 118, 125
Covalent bonds 3, 20, 22
Covalent radii 50
Crucial implication 149
Crystal 6, 7, 8, 9, 12, 16, 17, 18, 19, 21, 22, 23, 24
Crystal alterations 62
Crystal-chemical 65, 66, 76
Crystal-chemical method 65, 76
Crystal configurations 173
Crystal lattice 119
Crystalline graphite 182
Crystalline hydrates 151, 153
Crystalline inorganic compounds 70
Crystalline iodine 50
Crystalline powders 2
Crystallization 62
Crystallographic orientation 147
Crystal polarization 157
Cubic closest packing 172
Cubic face-centered arrangement 183
Cubic-shaped crystals 147
Cyclobutane 69

D

Decomposition 184
Dictates 2
Dihedral angle 174
Dislocations 2
Distribution 3, 4, 16
Disulfide bond 173
Donor-acceptor 11, 18, 28, 50, 52
Double bond 89, 91, 92, 100
Draw attention 121

E

Electrical arrangement 123
Electrical conductivity 116, 135
Electric arc 184, 187
Electric charges 10
Electric conductivity 184
Electron-attracting tendencies 128
Electron density 30
Electronegativity 91
Electronic interactions 17
Electronic polarization 155, 157
Electronic structure 10
Electron sea 132
Electron sharing 95, 96
Electron shell 14
Electrostatic attraction 12
Electrostatic forces 128
Environmental science 86
Equilibrium 102
Ethane molecule 99
Experimental dissociation energies 29
Experimental techniques 117

F

Face-centered molecule 62
First-order phase transitions 63
Freezing point 133
Fullerene molecule 183

G

Gaseous hydrocarbons 184
Gaseous samples 118
Geometric elements 124
Gibbs free energy 63
Glycerin compression 30
Gray shape polymeric chain molecules 62

H

Half fullerene sphere 184
Halogen atom serves 91
Heating 30
Heavy atoms 4
Heteroatomic distance 50
Hexagonal layer 173
Hexahalides 60
Homoatomic 29
Homologous series 86, 89
Homonuclear bonds 11
Hybridization 97, 106, 107
Hydrogen 93, 110, 111, 113
Hydrogen bond 47, 48, 54, 56
Hydroxyl functional group 91
Hydroxyl group 91, 92, 93, 94, 95

I

Infer information 75
Inorganic compounds 116, 135, 137
Interaction energy 28, 49
Interatomic distances 3, 9, 13
Interatomic interactions 12
Interatomic lengths 118
Inter-band optic changes 150
Intermolecular forces 86, 88
Internuclear distance 129
Iodine approximates 173

Ion-dipole 123
Ionic 10, 11, 12, 13, 16, 18, 19, 20, 21, 24
Ionic and covalent bonds 123
Ionic bond 124
Ionic Compounds Formation 126
Ionic connection 125
Isomer 100
Isotypism 9, 19, 25

K

Ketone group 92
Ketones 92
Kinetic factor significance 63

L

Lattice Energy 122
Lattice vibrations 157, 158
Lewis base 130
Lewis structure 95, 96, 97, 106
Linear function 73
Liquid 60, 66, 79, 80, 81
Liquid phosphorus 176
Liquid state elements 177
Liquid-vapor 133
Lorentz-Lorenz function 29

M

Macromolecular compounds 124
Macroscopic 61
Magnesium fluoride crystallizes 3
Manipulation 116
Materials science 86, 106
MD simulation configurations 67
Melting 60, 68, 69, 73, 75, 77, 79, 80, 81
Melting Point 133
Melting point temperature 173
Metal cation's oxidation 131
Metallic 4, 11, 13, 14, 16, 17, 18, 19, 20, 24
Metallic structures 181
Metal skeleton 17
Modern techniques 146
Modern technology 116
Molecular composition 101, 173
Molecular formula 86, 100
Molecular makeup 2, 3, 18
Molecular solids 71
Molecule rigid 99, 100
Molecules 2, 4, 5, 10, 11, 14, 15, 16, 17, 24
Molecule's internal covalent forces 124
Monoatomic ions 5
Monoclinic 148, 149
Mutual attraction 28, 50
Mutual repulsion 124

N

Natural state 176
Negative number 104
Neutral molecules 28, 50
Neutron diffraction (ND) 3, 47, 117
Neutrons 119
Nitrogen 93
Non-bonding electrons 14
Non-polar functional groups 91
Non-polar portion 89
Numerous polymorphic minerals 61

O

One-dimensional disorder 182
Optical activity 116
Optical analysis techniques 146
Optical anisotropy measurements 152
Optical axis 147, 148

Optical mono-axial crystals 149
Optical sign 149
Organic crystals 52, 54
Organic molecules 86, 88, 89, 93, 95, 96, 99, 101, 106, 107, 110, 113
Organic solvents 88
Orthorhombic symmetry 148

P

Pairing distribution functions 66
Peierls deformation 179
Phenyl group 94, 95
Phosphorus 176, 177, 178, 187, 191
Planes contrary 5
Planetary model 10
Polar compounds 122
Polarizabilities 30
Polyhedron 5, 6
Polyhedron face 5
Polymers 88
Polymorphic shape 62
Polymorphs 152, 163, 169
Polymorphs differ 61
Polyphosphides 177
Popular technique 117, 118
Positive charge 10
Potential curve 50
Potential energy 129
Potent ionic radii 65
Precise mathematical solution 123
Pressure reduce 172
Primary-polonium assembly 180
Primitive cells 121
Primitive hexagonal structure 180
Pyrolysis 184

Q

Quantum chemistry's 11

R

Random geometric structure 7
Reaction mechanisms 2
Real crystals 2, 117
Redistribution 96
Reflective index 151, 164
Refraction 30
Refractive index (RI) 134
Refractometry 146, 160, 165
Rhombohedral lattice 176
Room temperature 173, 176
Rotation ellipsoid 147
Rough approximation 13

S

Scattering 4
Schrodinger equation's 123
Second-order phase changes 63
Selenium atom 175
Selenium element 175
Selenium modifications 175, 187
Semiconductor crystals 153
Semiconductor elements 70
Semiconductor type 154
Separate chains encircle 175
Shock-treated quartz 153
Shock-wave compression 67
Sigma bond 99
Single crystals 2, 3, 158
Single electron pair 50
Single-walled nanotubes 184
Sodium chloride 126
Solid-liquid 133
Solid nitrogen-containing compounds 176
Solid state 55, 173, 175
Solid-state structures 3
Solubility 88, 107
Space group 8

Spatial arrangement 2, 116, 119, 123
Spatial configuration 61
Spatial distribution 3, 13, 18
Spectacular advancements 146
Spectroscopic method 150, 156
Spectroscopy utilizing electron energy transfer tests 151
Stereochemistry 2, 18
Stereoisomers 101
Structural chemistry 2, 11, 18, 20, 22, 24, 116
Sulfur exhibits 173
Surfaces touch 147
Synthetic and organic 86
Systematic nomenclature 3
Szigeti equation 157, 158

T

Temperature rises 2
Tetragonal shape 149
Thermal ellipsoid corresponds 3
Thermal ellipsoid size 118
Thermodynamic chemical 28
Thermodynamics 28
Thread-like macromolecules 124
Traditional polonium arrangement 178
Transition enthalpy 60
Transition entropy 60
Transverse lattice vibrations 157
Triple bonds 90, 99, 100

V

Valence-bond (VB) 3, 117
Valence electrons 15, 16, 17, 124, 128, 132, 182, 186
Valence shell 13, 14
Van der Waals 11, 15, 18
Van der Waals connections 28
VSEPR theory 96

W

Water molecule attraction 125
Water soluble substance 126
Wave functions 123
Wavelength 149, 151

X

X-ray diffraction 117, 118, 185
X-ray diffraction analysis 14

Z

zero-frequency maximum 151